Manufacturing Simulation with Plant Simulation and SimTalk

Preface

Based on the competition of international production networks, the pressure to increase the efficiency of production systems has increased significantly. In addition, the number of technical components in many products and as a consequence also the requirements for corresponding assembly processes and logistics processes increases. International logistics networks require corresponding logistics concepts.

These requirements can be managed only by using appropriate Digital Factory tools in the context of a product lifecycle management environment, which allows reusing data, supports an effective cooperation between different departments, and provides up-to-date and relevant data to every user who needs it.

Simulating the complete material flow including all relevant production, storage, and transport activities is recognized as a key component of the Digital Factory in the industry and as of today widely used and accepted. Cutting inventory and throughput time by 20–60% and enhancing the productivity of existing production facilities by 15–20% can be achieved in real-life projects.

The purpose of running simulations varies from strategic to tactical up to operational goals. From a strategic point of view, users answer questions like which factory in which country suits best to produce the next generation product taking into account factors like consequences for logistics, worker efficiency, downtimes, flexibility, storage costs, etc., looking at production strategies for the next years. In this context, users also evaluate the flexibility of the production system, e.g., for significant changes of production numbers — a topic which becomes more and more important. On a tactical level, simulation is executed for a time frame of 1–3 months in average to analyze required resources, optimize the sequence of orders, and lot sizes. For simulation on an operational level, data are imported about the current status of production equipment and the status of work in progress to execute a forward simulation till the end of the current shift. In this case, the purpose is to check if the target output for the shift will be reached and to evaluate emergency strategies in case of disruptions or capacities being not available unexpectedly.

In any case, users run simulation to take a decision about a new production system or evaluate an existing production system. Usually, the value of those systems is a significant factor for the company, so the users have to be sure that they take the right decision based on accurate numbers. There are several random processes in real production systems like technical availabilities, arrival times of assembly parts, process times of human activities, etc., so stochastic processes play an important role for throughput simulation. Therefore, Plant Simulation provides a whole range of easy-to-use tools to analyze models with stochastic processes, to calculate distributions for sample values, to manage simulation experiments, and to determine optimized system parameters.

Besides that, results of a simulation model depend on the quality of the input data and the accuracy of the model compared to the behavior of the real production system. As soon as assembly processes are involved, several transport systems with their transport controls, workers with multiple qualification profiles or storage logic, production processes become highly complex. Plant Simulation provides all necessary functionality to model, analyze, and maintain large and complex systems in an efficient way. Key features like object orientation and inheritance allow users to develop, exchange/reuse, and maintain their own objects and libraries to increase modeling efficiency. The unique Plant Simulation optimization capabilities support users to optimize multiple system parameters at once like the number of transporters, monorail carriers, buffer/storage capacities, etc., taking into account multiple evaluation criteria like reduced stock, increased utilization, increased throughput, etc.

Based on these accurate modeling capabilities and statistic analysis capabilities, typically an accuracy of at least 99% of the throughput values is achieved with Plant Simulation models in real-life projects depending on the level of detail. Based on the price of production equipment, a return on investment of the costs to introduce simulation is quite often already achieved after the first simulation project.

Visualizing the complete model in the Plant Simulation 3D environment allows an impressive 3D presentation of the system behavior. Logfiles can be used to visualize the simulation in a Virtual Reality (VR) environment. The support of a Siemens PLM Software unified 3D graphics engine and unified graphics format allows a common look-and-feel and easy access to 3D graphics which were created in other tools like digital product design or 3D factory layout design tools.

The modeling of complex logic always requires the usage of a programming language. Plant Simulation simplifies the need to work with programming language tremendously by supporting the user with templates, with an extensive examples collection and a professional debugging environment.

Compared to other simulation tools in the market, Plant Simulation supports a very flexible way of working with the model, e.g., by changing system parameters while the simulation is running.

This book provides the first comprehensive introduction to Plant Simulation. It supports new users of the software to get started quickly, provides an excellent introduction how to work with the embedded programming language SimTalk, and even helps advanced users with examples of typical modeling tasks. The book focuses on the basic knowledge required to execute simulation projects with Plant Simulation, which is an excellent starting point for real-life projects.

We wish you a lot of success with Tecnomatix Plant Simulation.

Dirk Molfenter †   November 2009
Siemens PLM Software

Table of Contents

Table of Examples

1 Introducing Factory Simulation

Simulation technology is an important tool for planning, implementing, and operating complex technical systems.

Several trends in the economy such as

- increasing product complexity and variety
- increasing quality demands in connection with high cost pressure
- increasing demands regarding flexibility
- shorter product life cycles
- shrinking lot sizes
- increasing competitive pressure

lead to shorter planning cycles. Simulation has found its place where simpler methods no longer provide useful results.

1.1 Uses

You can use simulation during planning, implementation, and operation of equipment. Possible questions can be:

- **Planning phase**
  Identification of bottlenecks in derivation of potential improvement
  Uncover hidden, unused potentials
  Minimum and maximum of utilization
  Juxtaposition of different planning alternatives
  Test of arguments regarding capacity, effectiveness of control, performance limits, bottlenecks, throughput speed, and volume of stocks
  Visualization of planning alternatives for decision making
- **Implementation phase**
  Performance tests
  Problem analysis, performance test on future requirements
  Simulation of exceptional system conditions and accidents
  Training new employees (e.g., incident management)
  Simulation of ramp up and cool-down behaviors
- **Operational phase**
  Testing of control alternatives
  Review of emergency strategies and accident programs
  Proof of quality assurance and fault management
  Dispatching of orders and determination of the probable delivery dates

S. Bangsow: Manufacturing Simulation with Plant Simulation, Simtalk, pp. 1 – 6, 2010.
© Springer Berlin Heidelberg 2010

1.2   Definitions

Simulation (source: VDI 3633)
Simulation is the reproduction of a real system with its dynamic processes in a model. The aim is to reach transferable findings for the reality. In a wider sense, simulation means preparing, implementing, and evaluating specific experiments with a simulation model.

System: (VDI 3633)
A system is defined as a separate set of components which are related to each other.

Model: A model is a simplified replica of a planned or real system with its processes in another system. It differs in important properties only within specified tolerance from the original.

Simulation run: (source: VDI 3633)
A simulation run is the image of the behavior of the system in the simulation model within a specified period.

Experiment: (source: VDI 3633)
An experiment is a targeted empirical study of the behavior of a model by repeated simulation runs with systematic variation of arguments.

1.3   Procedure of Simulation

According to VDI guideline 3633, the following approach is recommended:

1. Formulation of problems
2. Test of the simulation-worthiness
3. Formulation of targets
4. Data collection and data analysis
5. Modeling
6. Execute simulation runs
7. Result analysis and result interpretation
8. Documentation

1.3.1   Formulation of Problems

Together with the customer of the simulation, the simulation expert must formulate the requirements for the simulation. The result of the formulated problem should be a written agreement (e.g., a technical specification), which contains concrete problems which will be studied using simulation.

1.3.2   Test of the Simulation-Worthiness

To assess the simulation-worthiness you can, for example, examine:

- The lack of analytical mathematical models (for instance, many variables)
- High complexity, many factors to be considered
- Inaccurate data
- Gradual exploration of system limits
- Repeated use of the simulation model

1.3.3   Formulation of Targets

Each company aims at a system of targets. It usually consists of a top target (such as profitability), that splits into a variety of subtargets, which interact with each other. The definition of the target system is an important preparatory step. Frequent targets for simulations are for example:

- Minimize processing time
- Maximize utilization
- Minimize inventory
- Increase in-time delivery

All defined targets must be collected and analyzed statistically at the end of the simulation runs, which implies a certain required level of detail for the simulation model. As a result, they determinate the range of the simulation study.

1.3.4   Data Collection

The data required for the simulation study can be structured as follows:

- System load data
- Organizational data
- Technical data

The following overview is a small selection of data to be collected:

Technical data	
Factory structural data	Layout Means of production Transport functions Transport routes Areas Restrictions
Manufacturing data	Use time Performance data Capacity

Material flow data	Topology
	Conveyors
	Capacities
Accident data	Functional accidents
	Availability
Organizational data	
Working time organization	Break scheme
	Shift scheme
Resource allocation	Worker
	Machines
	Conveyors
Organization	Strategy
	Restrictions
	Incident management
System load data	
Product data	Working plans
	BOMs
Job data	Production orders
	Transportation orders
	Volumes
	Dates

1.3.5   Modeling

The modeling phase includes building and testing the simulation model. Modeling usually consists of two stages:

1   Derive an iconic model from the conceptual model.
2   Transfer the model into a software model.

1.3.5.1 First Modeling Stage

First, you have to develop a general understanding of the simulated system. Based on the objectives to be tested, you have to make decisions about the accuracy of the simulation. Based on the accuracy of the simulation, necessary decisions are taken about which aspects you want to simplify. The first modeling stage covers two activities:

- Analysis (breakdown)
- Abstraction (generalization)

Using the system analysis, the complexity of the system in accordance with the original investigation targets will be dissolved by meaningful dissection of the system into its elements. By abstraction, the amount of the specific system attributes will be decreased as far as it is practical to form an essential limited image of the original system. Typical methods of abstraction are reduction (elimination of not relevant details) and generalization (simplification of the essential details).

1.3.5.2 Second Modeling Stage

A simulation model will be built and tested. The result of modeling has to be included in the model documentation to make further changes of the simulation model possible. In practice, this step is often neglected, so that models due to the lack of documentation of functionality cannot be used. Therefore, there is a needing for commenting the models and the source code during programming. In this way the explanation of the functionality is still available after programming is finished.

1.3.6   Executing Simulation Runs

Depending on the objectives of the simulation study, the experiments based on a test plan will be realized. In the test plan, the individual experiments output data, arguments of the model, objectives, and expected results are determinated. It is also important to define a time span for the simulation experiments, based on the findings of the test runs. Computer runs spanning several hours or frequent repetitive experiments for the statistical coverage are not uncommon. In these cases it is helpful to check if it is possible to control the experiments by a separate programmed object (batch runs). The realization times for the experiments can be relocated partly in the night hours, so the available computing capacity can be utilized optimally. Input and output data and the underlying parameters of the simulation model must be documented for each experiment.

1.3.7   Result Analysis and Result Interpretation

The values, which will change in the modeled system, are derived from the simulation results. The correct interpretation of the simulation results significantly influences the success of a simulation study. If the results contradict the assumptions made, it is necessary to analyze what influences are responsible for the unexpected results. It is also important to realize that complex systems often have a ramp up phase. This phase may run differently in reality and in the simulation. Therefore, the results obtained during the ramp up phase are often not transferable to the modeled system and may have no influence for the evaluation (Exception: the ramp up phase of the original system has to be fully modeled).

1.3.8   Documentation

For the documentation of a simulation study, the form of a project report is recommended. The documentation should provide an overview of the timing of the study and document the work carried out. Of interest in this context is the documentation of failed system variants and constellations. The core of the project report should be a presentation of the simulation results based on the customer requirement specification. Resulting from the simulation study it makes sense to include proposals for actions in the documentation. Finally, we recommend describing the simulation model in its structure and its functionality.

2 Plant Simulation

2.1 First Steps

2.1.1 Online Tutorial

The online tutorial offers a quick start and guides you systematically in creating a simple simulation model. To start the tutorial, start Plant Simulation and left-click the tab INFO PAGES, then EXAMPLES, and TUTORIAL in the Explorer window.

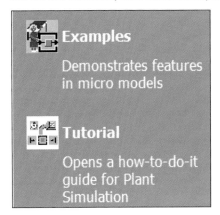

2.1.2 Examples

The sample model includes a variety of examples of small models that are thematically ordered and show how and with which settings you can use the components and functions. To open the models after starting Plant Simulation, click on the tab INFO PAGES, then EXAMPLES and EXAMPLES.

2.1.3 Help

The step by step help provides descriptions of steps, which are necessary to model several tasks. The step by step help is part of the online help, chapter "Step-by-step help".

The full functionality of version 9 is part of the online documentation. The manuals are available as Adobe Acrobat ® *. pdf files on the Plant Simulation installation CD and can be printed if required. The context-sensitive help in the dialogs of objects provides additional explanations of the dialog elements. To show context-sensitive help, click on the question mark on the top right corner of the dialog, and then click on the dialog element. The window of the context-sensitive help shows a reference to the corresponding SimTalk attribute at its end.

S. Bangsow: Manufacturing Simulation with Plant Simulation, Simtalk, pp. 7 – 15, 2010.
© Springer Berlin Heidelberg 2010

2.1.4   Website

Current information about Tecnomatix and Plant Simulation is available on the website http://www.plm.automation.siemens.com, or you use the direct link to the description of Plant Simulation:

http://www.plm.automation.siemens.com/en_us/products/tecnomatix/plant_design/plant_simulation.shtml.

2.2   Introductory Example

2.2.1   The Program

Start Plant Simulation by clicking on the icon in the program group or the desktop icon.

2.2.1.1 The Program Window

To define the layout, you can use the menu item: **VIEW – TOOLBOX**. Here you set what you see on the screen. A standard Plant Simulation window can, for example, contain the following elements:

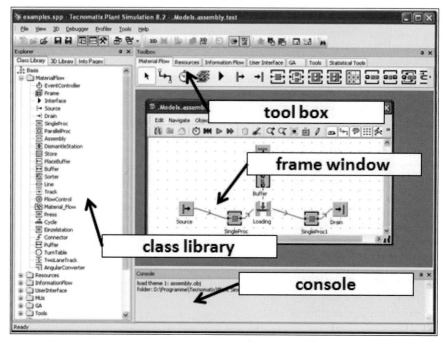

2.2.1.2 The Class Library

In the class library, you find all objects required for the simulation. You can create your own folders, derive and duplicate classes, create frames, or load objects from other simulation models.

To show the class library, you can use the command:

VIEW – VIEWERS – EXPLORER or the icon 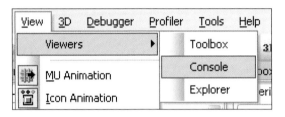.

You can hide the class library by clicking the X in the title bar.

2.2.1.3 The Console

The Console provides information during the simulation (e.g., error messages). You can use the command Print to output messages to the console. If you do not need the console, you can hide it by clicking the X in the title bar.

Show the console with the icon: 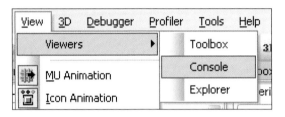 in the toolbar or in the menu with VIEW – VIEWERS – CONSOLE:

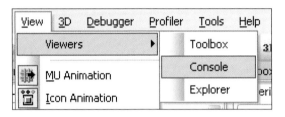

2.2.1.4 The Toolbox

The Toolbox provides quick access to the classes in the class library. You can easily create your own tabs in the toolbox and fill it with your own objects. The best way to show the toolbox is a click on the button 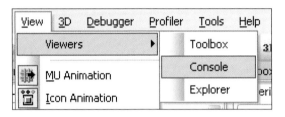.

2.2.2   First Simulation Example

2.2.2.1 Design of the Model

As a first example, a simple production line is to be build with a source (material producer), two workstations, and a drain (material consumer). Start Plant Simulation, and select the menu command:

FILE – NEW MODEL

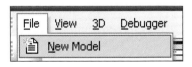

It opens the Plant Simulation class library with the basic objects and a Frame (window). Simulation models are created in the object Frame.

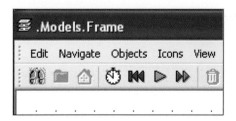

2.2.2.2 Insert Objects into the Frame

To insert objects into the Frame, you have two options:

- Click on the object icon in the toolbox, and then click in the Frame. The object will be inserted into the Frame at the position, at which you clicked.
- Another way: Drag the object from the class library to the Frame and drop it there (drag and drop).

Insert the following objects in the Frame:

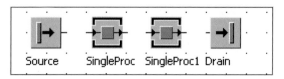

2.2.2.3 Connect the Objects

You have to connect the objects along the material flow, so that the different parts can be transported from one object to the next. This is what the object Connector does. The Connector has the following icon in the toolbar:

Click the Connector in the toolbar, then the object in the Frame, which you want to connect (the cursor changed its icon on Connector) – click the next object … If you want to insert several Connectors successively, hold down the CTRL key. Result:

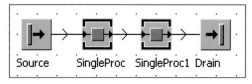

2.2.2.4 Define the Settings of the Objects

You have to define some settings in the objects such as processing times, capacity, information for setup, failures, breaks, etc. Properties can be easily set in the dialogs of the objects. You can open its dialog by double-clicking an object.

Example 1: Properties of the SingleProc

Set the following values: SingleProc stations: processing time: 2 minutes, Drain: Processing time: zero seconds, Source: 2-minute interval

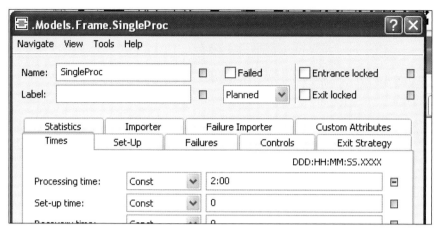

The APPLY button saves the values, but the dialog remains open. OK saves the values and closes the dialog. Finally, you need to insert an event controller. It coordinates the processes that run during a simulation.

Click on the event controller [symbol] in the toolbox, then in the Frame.

2.2.2.5 Run the Simulation

Open the control panel by double-clicking on the icon of the event controller.

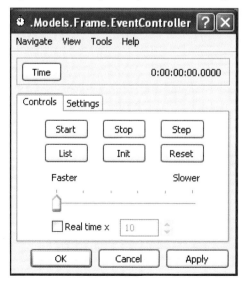

Click the **START** button to start the simulation, and **STOP** to stop the simulation. In the Frame, the material movements are graphically displayed. You can now change the model to see what happens…

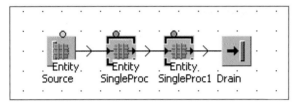

2.3 Modeling

2.3.1 Object-Related Modeling

In general, only a limited selection of objects is available for representing the real installation. They talk, e.g., of model objects, which show the real system with all the properties to be investigated. Hierarchically structured system models are best designed top-down. In this way, the real system will be decomposed into separate functional units (subsystems). If you are not able to model sufficiently precise with the available model objects, you should continue to decompose, etc.

Each object must be described precisely:

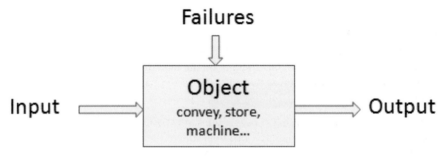

The individual objects and the operations within the objects are linked to an overall process. This creates a Frame. With the objects and the Frame various logistical systems can be modeled.

2.3.2 Object-Oriented Modeling

2.3.2.1 Objects and Properties

The hierarchical structure of an object allows to be exactly addressed (analogous to a file path). A robot Rob1 may be addressed in the hierarchy (the levels are separated by a period) as follows:

production1.press_hall.section1.cell1.Rob1

Rob1 itself is described by a number of properties such as: type of handling, speed, capacity, lead times, etc. All properties, which describe Rob1, are called "object". An object is identified by its name (Rob1) and its path:

(production1.press_hall.section1.cell1.Rob1)

The properties are called attributes. They consist of a property description (attribute type), for example Engine type and a property value (attribute value) for instance: HANUK-ZsR1234578.

2.3.2.2 Classes and Instances

In object-oriented programming, a class is defined as follows:

A class is a user-defined data type. It designs a new data type to create a definition of a concept that has no direct counterpart in the fundamental data types.

Example: You want to create a new type of transport unit that cannot be defined by standard types. All definitions (properties, methods, behavior), required for creating a new type, are called a "class".

The individual manifestation of the class is called an instance of the class (e.g., Transport – Forklift (general); Instance: Forklift 12/345 (concrete)). The instance has the same basic properties as the class and some special characteristics (such as a specific name).

2.3.2.3 Inheritance

In Plant Simulation, you can create a new class based on an existing class (derive of class, create a subclass). The original class is called base class. The derived class is called subclass. You can expand a data type through the derivation of a class without having to redefine it. You can use the basic objects of the class by employing inheritance.

Example: You have several machines of the same type; most of the properties are the same. Instead of defining each machine individually, you can define a basic machine. All other machines are derived from this basic machine. The subclasses inherited the properties of the base class they apply to these classes as if they were defined there.

2.3.2.4 Duplication and Derivation

Example 2: Inheritance 1

Select the SingleProc in the class library. Click the right mouse button to open the context menu. Select DUPLICATE from the context menu.

sma		
ore	Duplicate	Ctrl+Drag
ace	Derive	Shift+Ctrl+Drag

Plant Simulation names the duplicate SingleProc1. Change the processing time in the class SingleProc to 2 minutes. Open the dialog of the SingleProc1. The processing time has not changed.

The duplicate contains all the attributes of the original, but there is no connection between the original and the duplicate (there is no inheritance). You can also create duplicates using the mouse: Press the Control key and drag the object to its destination, then drop it.

Example 3: Inheritance 2

*Now do the same with **DERIVE**. Select the SingleProc again, click the right mouse button and select **DERIVE** from the context menu.*

*Plant Simulation names the new class SingleProc2. With **DERIVE** you created an instance of the class. This instance can either be a new class (in the library) or an object (e.g., in a Frame object). Initially, the instance inherits all the characteristics of the original class. Now, change the processing time of the SingleProc-class to 10 minutes (10:00). Save the changes in the SingleProc, and open the dialog of SingleProc2. SingleProc2 has applied the change of the processing time of the SingleProc-class.*

You can also derive in the class-library with CTRL + SHIFT and dragging the mouse. You can navigate to the original class from an object or from a derived class. Double-click the class/the object then select:

NAVIGATE – OPEN ORIGIN

It will open the dialog of the original class. If you drag a class from the class library into a Frame object, the new object is derived.

Example 4: Inheritance 3

Open a Frame object. Add a SingleProc to the Frame object (via drag and drop from the library). Change the processing time of the SingleProc in the library and check the processing time in the Frame. The values are inherited from the object in the library.

Duplicating Objects in the Frame

To duplicate an object in the Frame, hold down the Ctrl key and drag the object to a free spot on the Frame (the mouse pointer shows a "+"). With the object you can also activate inheritance to the initial class.

Test: Change the processing time of the SingleProc in the library: The processing time of both SingleProcs in the Frame also changes. If you change the processing time of one SingleProc in the Frame, the processing time of the other SingleProc in the Frame will not be changed (the duplicate has no inheritance relationship with its original, but with the original class of the original). In the object dialogs you can easily identify which values are inherited and which have been entered in the instance. Each attribute shows a green toggle button to the right. This is green, if the value is inherited, and yellow with a minus sign inside if the values are not the same as in the original class.

Value inherited	Value changed (inserted)
1:00 □	5:00 ⊟

To restore inheritance of a value, click the button after the value and then click APPLY.

3 Standard Classes in PLANT SIMULATION

3.1 Overview

The standard classes can be classified into six categories:

1. Material flow objects
2. Resources
3. General objects
4. Mobile objects
5. Lists and tables
6. Display objects

3.2 Material Flow Objects

Mobile and static material flow objects are the basic objects of a model. The mobile units (transporters, containers, parts) represent the physical or logical objects, which move through a model. These units are transported through the simulation model by the active or passive material flow objects (e.g., a part is located on a conveyor; the passive part will be transported through the model by the active material flow object "conveyor"). Active objects are SingleProc, ParallelProc, AssemblyStation, DismantleStation, Line, TurnTable, AngularConverter, Sorter, and Buffer. They actively transport the mobile material flow objects along the connectors. Source and drain are used to create and destruct mobile objects (MUs). Thus, they represent the model borders. Passive material flow objects are Store, Track, and TwolaneTrack. These objects do not pass on the MUs automatically. The FlowControl object (which itself cannot store MUs) represents junction or distribution strategies.

3.2.1 General Behavior of the Material Flow Objects

Active material flow objects can receive mobile objects, store them for a certain amount of time, and then automatically pass them on to the next object. Passive material flow objects cannot automatically pass on MUs (Example: a MU will remain in storage until it is removed by a method from the object). The passive object "track" can be only used in a meaningful way together with the object transporter, which means that this MU moves on the track with a certain speed.

S. Bangsow: Manufacturing Simulation with Plant Simulation, Simtalk, pp. 17 – 74, 2010.
© Springer Berlin Heidelberg 2010

3.2.1.1 Time Consumption
Example 5: Material Flow – Time Consumption

To demonstrate a small example, model the following Frame:

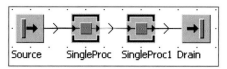

An object receives MUs when it has free capacity and is neither failed nor paused. If one of the conditions is not met or the gate is closed because of the recovery, or cycle times, the object rejects the MUs. The moving object will be entered at the end of a blocking list. Once the object can receive MUs again, the first MU in the blocking list will be moved and processed (First in First Out).

Example: Open the source by double-clicks on the object. The source generates MUs (entities by default). The interval between the individual MUs can be set here. The default is 0 minutes. Change the interval to 1 minute:

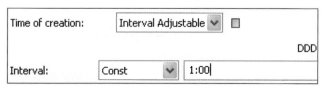

Set the processing time of the next object (SingleProc1) to 2 seconds. Double-click on the SingleProc1 in the Frame. In the dialog select the tab TIMES.

Enter the time into the field Processing time (enter a 2, and confirm with Apply). If the following station has a processing time of more than a minute, the MUs should accumulate. Set the processing time of SingleProc2 to 2 minutes (2:00). Let the simulation run for a while. Stop the simulation (Stop in the Eventcontroller) and then open the SingleProc1 (double-click). Select the tab STATISTICS.

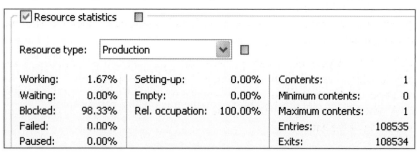

The station is blocked most of the time by the successor, it could pass on the MUs, and the following station (which is still processing) cannot yet store the MUs. The time passed is called blocked time.

Entrance Gates

At the entrance of an object, there are two "gates":
1. The object is empty, not failed or paused.
2. An entire multiple of a cycle time.

The cycle time is used for synchronization. Although the previous station is ready earlier, the part waits until the cycle is over. Only then, the part will be transferred to the next object. The processing duration of a part on a station consists of three parts:

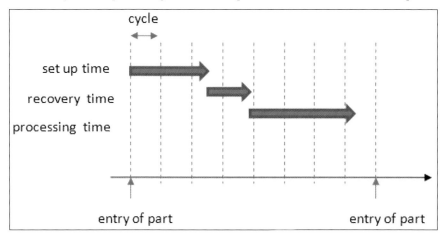

SETUP TIME: The Setup Time is the time which is required to set up a basic object for processing another type. The type is determined by the name of a MU (MUs with the same name have the same type). You can also use a setup time after a certain number of parts have been set up, for example, for regular tool changes.

RECOVERY TIME: At the entrance of a basic object, there is a gate, which closes for a specified time after an MU entered. This way you model robots, which require a certain time to insert parts into the machine.

PROCESSING TIME: The processing time determines how long an MU stays on the object after the setup time, before Plant Simulation tries to move the MU to a succeeding object.

CYCLE TIME: Cycle times can be used for synchronizing productions. They specify in which interval or in which integral multiple of an interval entering the workstation is possible (e.g., every 32.4 seconds). The gate opens every 32.4 seconds. A new MU can only enter after the end of the previous processing and the next opening of the gate.

3.2.1.2 Capacity

The capacity determines how many MUs can be located on the object at the same time. If the limit of the capacity is reached, MUs will no longer be transferred. The dimension of MUs is usually not taken into account. Only the objects: Track, Line, AngularConverter, and Turntable take the length of the MUs into account (attribute length).

3.2.1.3 Blocking

MUs, who want to enter into a full object, will be rejected (and entered into a blocking list). If the object has several successors, the transfer request is made consecutively to the following objects. The MU will then be transferred to the next free object.

Example 6: Blocking

Create the following Frame. Processing times: Mach1 and Mach4 every 2:30 min, Mach2 und Mach3: 5 min, the source produces one part every 2:30 min.

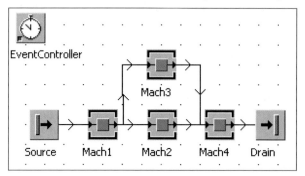

Block Mach2 (check the box Failed, then click Apply) and look, what happens!

You can change the behavior of the distribution of Mach1 on the tab **EXIT STRATEGY**. It provides a number of strategies:

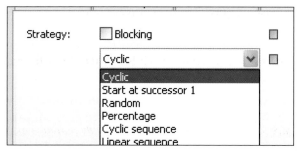

Blocking here means: If you select **BLOCKING**, and a transfer request could not be met, the transfer request will be entered in the blocking list of the object. It will wait, until the next object can receive MUs again. When blocking is not selected, the MU will be transferred to another free successor.

The objects provide the following types of blocking:

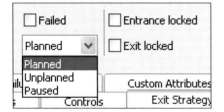

Failed: The object cannot receive MUs. Already finished MUs are moved on. As long as the object is failed, the setup or processing time is paused until the end of the failure. (This way you can simulate, what happens when a machine fails.) Failed objects are marked with a red LED.

Paused: An object can be paused (blue LED). The processing or setup time is interrupted. Already finished MUs will be moved on. The device cannot receive MUs. The same effect has Unplanned (look shift-calendar). By unplanned, you can simulate times outside the working time. The reset button in the Eventcontroller resets any blockages. Pauses must be manually reset (clear the option Pause).

Entrance locked: It is not possible to move MUs; the MUs will be entered in the blocking list. After the end of the failure, the MUs will be processed.

3.2.1.4 Failures

To achieve a most realistic simulation, you have to include some events which disrupt the normal flow of materials. You can position these events accurately or randomly. You can take into account times for retooling and maintenance times, accidents, machine failures, and others.

There are two possiblities to model failures:

- Using statistical distributions
- Using the Mean Time To Repair (MTTR) and the availability

Define Failures
The dialogs of the material flow objects provide the tab FAILURES: Clear the option AVAILABILITY.

Example 7: Failure 1

A machine needs maintenance every 1,000 operating-hours with duration of 3 hours. It requires these settings:

Statistics	Importer	Failure Importer		Custom Attributes
Times	Set-Up	Failures	Controls	Exit Strategy

☑ Active   ☐

DDD:HH:MM:SS.XXXX

Start:	Const ∨	41:16:00:00	☐
Stop:	Const ∨	0	☐
Interval:	Const ∨	41:16:00:00	☐
Duration:	Const ∨	3:00:00	☐

☐ Availability   ☐   Failure mode relates to: ProcessingTime ∨ ☐

Active: The check box turns all kinds of failure events on or off.

Start: With "Start", you can define the beginning of the failure. You can also se-lect a statistical distribution.

Stop: End of failure

Interval: Enter an interval between the end of the last failure and the beginning of the next failure (trouble-free time). If the value of the selected interval is zero and the value of the selected duration is greater than zero, then one single failure occurs.

Duration: Duration of the failure (duration = 0 means no failure)

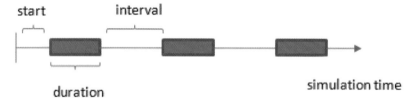

Failure mode relates to simulation time: For this setting, Plant Simulation con-sumes the time you have entered as interval, independent of the state of the object (paused or operational).

Failure mode relates to processing time: For this setting, Plant Simulation con-sumes the time you have entered as interval while the object is working (not paused, waiting, or unplanned).

Failure mode relates to operating time: For this setting, Plant Simulation consumes the time you have entered as interval, if the object is not paused (working or waits).

Example 8: Failure 2

A machine (3-shift-mode) needs maintenance lasting 1.5 hours every 22.5 hours of processing time. The succeeding machine requires 30 minutes maintenance every

3.5 hours of processing time. Both machines have a processing time of 2 minutes. To ensure a smooth material flow, a buffer is located between the machine1 and machine2. How many places must the Buffer have?
   Create the following Frame:

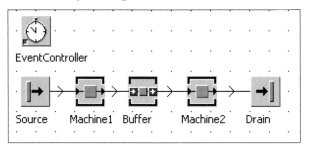

Let the simulation run for 2 days to observe the buffer. The line should not jam; the buffer should not be oversized either. You can make an evaluation of the necessary buffer size using the tab statistics of the object buffer.

Availability (MTTR, MTBF)
You can specify the availability and an average repair time for the processing stations. The system then calculates a mean time between failures (Mean Time between Failures, MTBF). The duration of outages and failure-free times will be randomly distributed. You have to specify a random number stream (a random number stream is a series of random numbers). The availability will be calculated using the following formula:

   Availability = MTBF/(MTBF+MTTR).

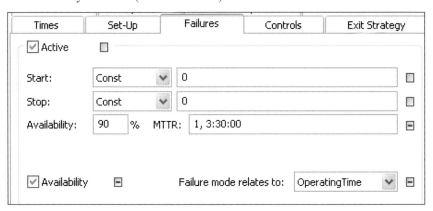

Enter a number between 0 and 100% for the **AVAILABILITY**. The availability is the probability that a machine is ready to use at any time. The availability is based on a combination of MTBF[1] and MTTR[2]. The duration of the failure as well as the dis-

---

[1] MTBF—Mean time between failures.

tance between the failures are randomly distributed. Plant Simulation selects the Erlang distribution for the duration and the Negexp distribution for the interval.

Starting with Plant Simulation version 9, you can create a series of failures for an object. This way, you can more realistically model the failure behavior of machines and plants. You can, for example, set maintenance intervals and tool changes of a machine in one dialog and without programming failures.

Example 9: Multiple Failures

We will use a machine with the following failure behavior: For each 5 hours of processing time, a tool change takes place which takes 30 minutes. Every 1000 hours of operating time, regular maintenance taking 2 hours takes place, 5% loss (random) relative to the operating time, 2 hours MTTR. To consider these values in the simulation, you need to proceed as follows in Plant Simulation version 9. Click the tab **FAILURES**, *click the button* **NEW**:

Enter the failure data into the dialog, e.g., tool change:

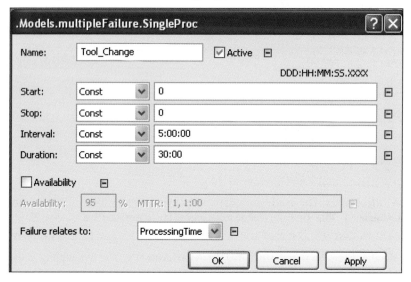

---

[2] MTTR—Mean time to repair.

All failures are displayed in a list. Double-clicking an item in the list allows you to edit the individual failures.

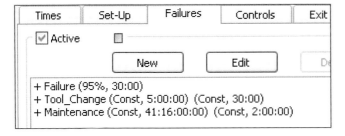

3.2.2   The Source

3.2.2.1 Basic Behavior

The source creates mobile objects (MUs) according to your definition. The source can produce different types of parts in a row or in mixed order. For defining batches and determining the points in time, the program provides different methods. The source as an active object tries to transfer the produced MU to the connected successor.

3.2.2.2 Settings

Mode: Mode determines how to proceed with MUs, which cannot be transferred.

Blocking means that the generated MUs will be saved (it will produce no new MUs). If you select "Non-blocking", the Source creates another MU exclusively at the time of creation you entered.

Time of creation:

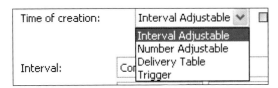

Interval Adjustable: The production dates are determined by three figures: start, stop, and interval. The first part is produced at the time "Start". Other parts are produced at an interval. The production of the parts ends with stop. You can enter statistical distributions for all three values.

Time of creation:	Interval Adjustable ✔ ☐		
			DDD:HH:MM:SS.XXXX
Interval:	Const ✔	1:00	⊟
Start:	Const ✔	0	☐
Stop:	Const ✔	0	☐

Number Adjustable: Number and interval (a certain number at specified interval) determine the production dates.

Time of creation:	Number Adjustable ✔	⊟	Amount:	10	⊟
				DDD:HH:MM:SS.XXXX	
Creation times:	Const ✔	10:00			⊟

The settings above will produce 10 parts after 10 minutes simulation time only once.

Delivery Table: The production times and type of parts to be produced are taken from a table (delivery table). Each line in the delivery table contains a production order. For this purpose, you have to add a table to your Frame.

Example 10: Source Delivery Table

Create the following Frame. You can change the length of the line by draging the corner points at the right-hand side.

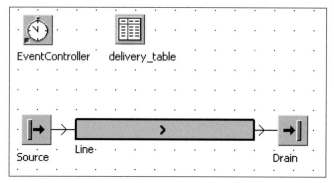

Duplicate the object Entity three times in the class library. Rename these dupli-cates to part1, part2, and part3. The source should produce five parts of type part1 after 2 minutes, two parts of type part2 after 10 minutes, and four parts of type part3 after 15 minutes. Select TIME OF CREATION – DELIVERY TABLE in the dialog of the source. Next, click in the lower part of the window on the button with the three points. Select the table in the following dialog.

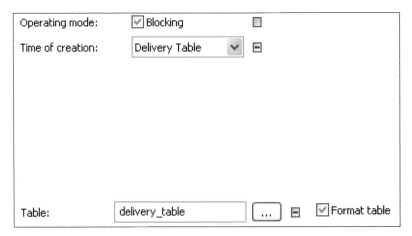

Finally, click OK. The name of the table will be entered into the dialog of the source. Now open the table in the Frame by double-clicks, and enter the following:

	time 1	object 2	integer 3	stri 4
string	Delivery Time	MU	Number	Nar
1	2:00.0000	.MUs.part1	5	
2	10:00.0000	.MUs.part2	2	
3	15:00.0000	.MUs.part3	4	

The source now produces parts as specified in the table. After the last part has passed the drain, the simulation will be finished.

MU-Selection: The following settings are available:

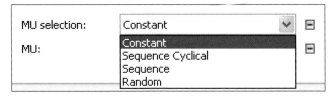

Constant: Only one MU-type will be produced. Select the path to the respective MU in the dialog (Object Explorer).

Sequence Cyclical: Parts are produced according to the sequence you entered in a table (see delivery table). Enter the path to the table into the text box. If the check box "GENERATE AS BATCH" is checked, the quantity will be produced at one time. When the sequence is completely produced, the production sequence will repeat.

Sequence: See Sequene cyclical; after the end of processing the entries no repetition will take place.

Random: The production is based on a random table.

Example 11: Randomly Produce MUs

*Part1, Part2, and Part3 are to be produced. The ratio is 30% for Part1, 60% for Part2, and 10% for Part3. First, create the parts in the class library. Try to change the color of the parts (right mouse button – **EDIT ICONS** ...).*
 Create the following Frame:

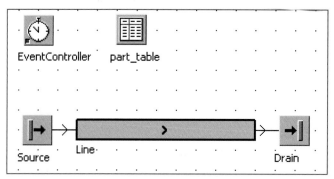

Now select the following settings in the source:

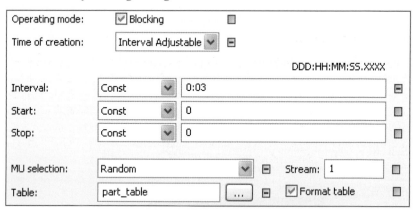

Confirm your changes with Apply or Ok.

Open the table by double-clicking it. Enter the following data:

	object 1	real 2	string 3
string	MU	Frequencies	Name
1	.MUs.part1	0.30	
2	.MUs.part2	0.60	
3	.MUs.part3	0.10	

Now start the simulation. The parts will be produced in a "mixed" order.

3.2.3   The Drain

The Drain is an active material flow object. The Drain has a single place and destroys MUs after processing them. Set the processing time to 0 second in the Drain or to the time a following process would require. The Drain collects a number of important statistical data such as throughput, number of destroyed parts, etc. Click on the tab TYPE STATISTICS.

Times	Set-Up	Failures	Controls	Statistics	Type Statistics	Custom Attributes

☑ Type dependent statistics   ☐

[Detailed Statistics Table]

Working:	100.00%	Average lifespan:	9.0000
Delayed:	0.00%	Average exit interval:	3.0000
Setup:	0.00%	Total throughput:	113799
Failed:	0.00%	Throughput per hour:	1200.00
Paused:	0.00%	Throughput per day:	28800.00

3.2.4   The SingleProc

The SingleProc accepts exactly one MU from its predecessor. After the setup, recovery, and processing time, the MU will be transferred to one of its successors. While an MU is located on the object, all other newly arriving MUs will be blocked. Only after a successor is free and not occupied, the MU will be transferred (it is possible to define different transfer procedures). You can use it to simulate all machines and jobs, which handle one part after another. One part at a time can be located on the workplace or the machine.

3.2.5   The ParallelProc

3.2.5.1 Basic Behavior and Use

The basic behavior of the ParallelProc is the same as that of a SingleProc with multiple places. Without a control, a newly arriving MU will always be placed on the place, which was empty for the longest time. When an MU with a different name arrives, the entire object will be set up.

Example 12: ParallelProc

After deburring, parts will be treating in a coloring. Because deburring has a long processing time, there are several places for deburring. Source1 delivers one part every 2 seconds. The deburring station has five places with a processing time of 10 seconds per place and part. Coloring takes 2 seconds. The Frame with SingleProcs could look as follows:

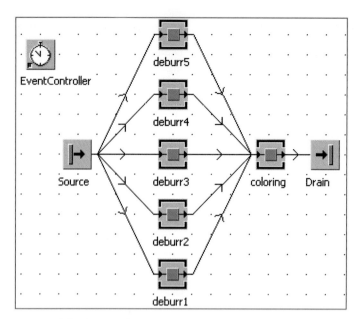

To simplify the simulation, you can use a station with five processing stations (ParallelProc).

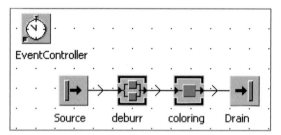

3.2.5.2 Settings

The processing stations of a parallel station are arranged in a matrix (tab **ATTRIBUTES**).

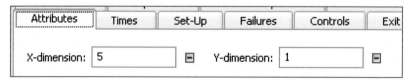

Every "row" takes on x places (x-dimension) in y "columns" (y-dimension). The number of places results from the multiplication of x-Dimension and y-Dimension. If you want to reduce the dimension of the ParallelProc, then it may not find MUs on the places (otherwise, you get an error).

Normally, each place of a ParallelProc has the same processing time. Neverthe-
less, it is possible to define different times for each place.
Proceed therefore like this:

1. Define the number of places (x-dimension, y-dimension)
2. Insert a table in the Frame (folder InformationFlow).
   Select in the tab "TIME" in the list "PROCESSING TIME": List (place)

3. Enter the name of the table into the field.
4. Enter the processing times for the individual stations into the table (analogous to the position of the places in x- and y-dimension).

Note:
Until Plant Simulation 8.2 you need to format the table. You have to allocate according to the x-dimension of the ParallelProc a number of columns the data type
time. From version 9, Plant Simulation formats the table according to the dimension of your ParallelProc (x columns and y rows) if the data type of the table columns does not match. Because of that, you have to set the x-dimension and y-
dimension values for the ParallelProc before assigning the table.

Example 13: ParallelProc; Different Processing Times
A production line has four deburring stations with different technical equipment.
For this reason, the individual stations have different processing times: station 1
and station 4 has one minute each, station 2 two minutes and station 3 four minutes.
Create the following Frame:

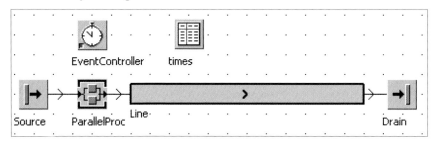

Settings: Source: interval 20 seconds, Line: length 12 meters, speed 0.08 m/s;
Drain: 0 seconds processing time. To define the processing times, follow these steps:

Set the dimension of the ParallelProc.

*Select "**LIST (PLACE)**" from the list **PROCESSING TIME**. Enter the name of the table "times" into the text box (or drag the table from the Frame in the field).*

Attributes	Times	Set-Up	Failures	Controls	Exit

Table[tim

Processing time:	List(Place) ∨	.Models.Frame.times

Confirm your changes by clicking OK.
Open the table and enter the times.

	time 1
1	2:00.0000
2	4:00.0000
3	1:00.0000
4	2:00.0000

3.2.6   The AssemblyStation

The AssemblyStation adds mounting parts to a main part or simulates assembly processes by destroying the single parts and generating the assembled part. The AssemblyStation facilitates the simulation of assembly operations.

Example 14: Assembly

Before coloring a part it will be mounted onto a support frame. The coloring is not possible without using the support frame. The assembly lasts 2 minutes, coloring also takes 2 minutes. Main part: support_Frame (Container), mounting-part: part (Entity). Create the following Frame:

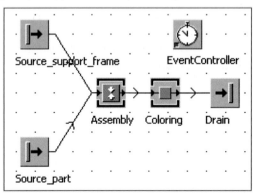

*Make sure that you first connect the source_support_frame, then source_part with the assembly. You have to convince the support_source_frame to generate support_ Frames (default is Entity). Double-click on the source and select **MU Container**.*

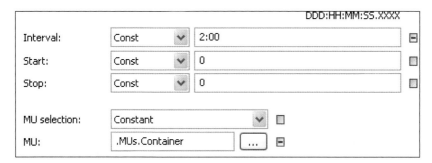

The object AssemblyStation has the following attributes:

Assembly table with: *Select the parts, which you want to assemble, according to different points of view:*

- *MU types*
- *Predecessor Number*

If you do not select an assembly list, one of each part will be assembled.

Select **Predecessors** *and open the predecessor-table (button open): Enter the number of the predecessor and the amount of assembled parts into the list. If you select the assembly mode "Attach MUs", you should not enter the main part into the list. In the example above, one part of predecessor 2 is to be mounted: Select Assembly table with – Predecessor, and then click Open. Enter the following to the list:*

	Predecessor	Number
1	2	1

If you select MU-types, enter the name of the MU-class and the respective number of parts into a table.

Note:
You can show the numbers of the predecessors. Select **VIEW – OPTIONS – SHOW PREDECESSORS** in the Frame window.

The number of the predecessor will be displayed on the connector.

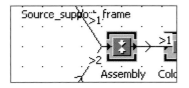

Main MU from predecessor: Here, you define which workstation provides the main part. The number is derived from the sequence in which you established the connections. Please note that the main MU itself must be able to accept parts (e.g., container), if you select the option ATTACH MUS.

Assembly mode: You can "load" parts on a main part (the main part must have sufficient capacity for example as container) or destroy all parts and create a new part (assembly part).

Exiting MU: The main part (with the loaded components) or a new part can be moved from the assembly. If you create a new part, you have to select it.

3.2.7   The Buffer

Plant Simulation distinguishes between two types of buffers:

(a) PlaceBuffer
The MUs pass the PlaceBuffer one after another in the "processing time." MUs cannot pass each other within the buffer. Only when the MU has reached the place with the highest number, it can be passed on. When the last MU has been passed on, all other MUs can move forward one place. The processing time can be specified only in relation to the entire buffer (e.g., dwell time in the buffer 20 min), not in relation to a single place (ten places, 2 minutes). The attribute ACCUMULATING determines whether the exit of the buffer is blocked (e.g., the successor is occupied) and any following MUs move up (Accumulating= TRUE) or have to wait.

(b) Buffer
The buffer does not have a place-oriented structure. After the processing time is over, you can remove the MU again. You can determine a mode for unloading:

- Buffer type Queue: First in First out
- Buffer type Stack: Last in First out

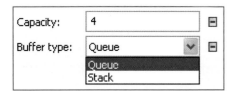

Settings:
Capacity: number of places in the Buffer; enter -1 for an infinite capacity

Capacity:	4	⊟
☑ Accumulating		⊟

Times: Processing time (dwell time of a part in the buffer), recovery time, cycle time

Processing time:	10:00	⊟

3.2.8 The DismantleStation

3.2.8.1 Basic Behavior

The DismantleStation dismantles added parts from the main part or creates new parts. It facilitates modeling dismantling operations.

Example 15: DismantleStation

A machine is loaded with a palette (12 parts). The machine unloads the parts at a fixed position with an internal loader from the palette and loads the parts into the machine. After completion, the parts are stored on a different palette at a second position. The palettes are transported to the machine on a three-meter-long conveyor. Behind the machine another three meter conveyor for palettes with finished parts is available. Unloading of the parts can be easily realized with a DismantleStation (the parts are unloaded onto the machine; the empty palettes are transferred to the place where the finished parts will be loaded). Loading the parts on the palette can be realized with the AssemblyStation. Create a part (entity) and a palette (container) in the class library. Enter a capacity of 12 (x-Dimension: 3, y-Dimension: 4, length and width each 0.5 meters) into the dialog of the container.

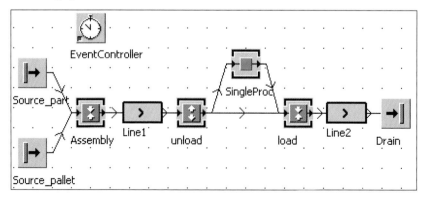

Settings: Source_part: it will generate entities (part), interval: 1:05, Source_pallet: it will generate containers (pallet), interval: 12:00, AssemblyStation: processing time: 0, main part: pallet; assembly mode: attach MUs; twelve parts from predecessor 2 (connect first Source_pallet, then Source_part with the assembly station), line1: 3 meters, speed 1m/s, line2: 3 meters, speed 1 m/s, machine: 1 min processing time, loading: main part from unloading (connect at first), 4 seconds processing time, assembly mode attach MUs, twelve parts from machine. To start the simulation, an empty palette must be ready at the loading station. The easiest way is to create an empty palette on line1. The DismantleStation reaches for the empty palette next to the loading station. So that the assembly station receives the palette on the right connector, we need a method object (folder InformationFlow). Rename the method object to INIT (this method is called when you click INIT in the Eventcontroller). Double-clicking the object opens an editor. If the font in the editor is grayed out and you cannot enter your source code, you have to first deactivate inheritance by clicking the button:

Drag your palette class from the class library to the editor (between do and end). The absolute path of the class will be entered into the editor. Complete the source code like this:

```
is
do
      .MUs.pallet.create(line1);
end;
```

Then close the window and save your changes.

Dismantle Sequence
Select how the DismantleStation distributes MUs to its successors from the drop-down list sequence.

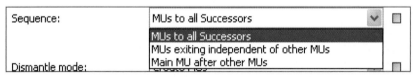

MUs to all Successors: If you have chosen the option Create MUs in the drop-down list Dismantle mode, the DismantleStation creates a new MU for each successor and transfers it to the successor. If you have chosen MUs detach from the drop-down list Dismantle mode, then the DismantleStation distributes the MUs round to its successors. Please note that the DismantleStation transfers the main part to the successor with the number you have entered into the field "Main successor to MU".

For the following three menu commands, Plant Simulation requires entries into the dismantle list.

MUs exiting independent of other MUs: Each MU is trying to move as soon as possible to the given successor.

Main MU after other MUs: The mounted parts exit the DismantleStation first followed by the main part.

This setting is important if you simulate unloading parts. The empty palette exits empty at the end the dismantle station and not before, for example if the individual parts cannot be delivered to the successor fast enough. To illustrate this, take a look at the following small example.

Example 16: Dismantle Station, Exit Sequence

Palettes have to be loaded and unloaded. The parts are weighed after unloading and then destroyed by a drain. Weighing takes 5 seconds. Create the following Frame:

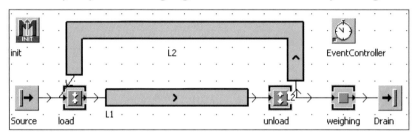

Settings: The source creates one part every 4 seconds (Entity). First Connect L2 then the Source with the station load. Set the capacity of the container in the class library to 10. Set the assembly station so that 10 parts will be loaded onto the main part (from L2). Assembly and dismantling take no time (processing time 0 seconds). L1 and L2 each have a speed of 1 m/s. First connect the station unload with L2 and then unload with weighing. The init-method should create two containers on the line L2. Complete the init method as follows.

```
is
do
    .MUs.Container.create(L2,2);
    .MUs.Container.create(L2,4)
end;
```

Select the following settings in the DismantleStation (unload):

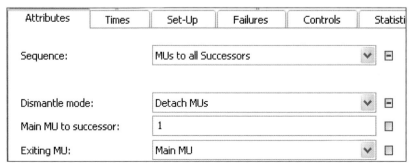

The palette exits the dismantle station before the last parts. In most cases, such behavior does not fit reality. The container must exit the dismantling station after the single parts. To do this, change the settings of the dismantling station:

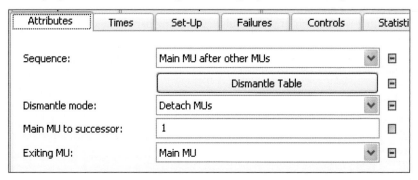

In the dismantle table you have to define how many of which parts are to be transferred to any successor.

	MU	Number	Successor
1	.MUs.Entity	10	2

Now all parts will be unloaded first, before the main part will be transferred.

Dismantle Mode
Select how Plant Simulation deals with mounted parts.
   It provides two modes:

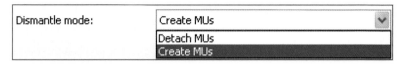

Detach MUs: Plant Simulation unloads the mounted parts from the main part and transfers them to the successors, which are contained in the dismantle table.
   Create MUs: The DismantleStation creates new parts.

Main MU to Successor
This successor number may not be contained in the dismantle table (error message).

Exiting MU
Here, you determine how Plant Simulation handles the main parts.

3.2.8.2 Cycle

With the cycle object, you can synchronize a set of objects. The MUs will be passed on only if all stations in the balanced line are finished, neither failed nor paused, and the next station is ready to receive MUs. Connectors must link the stations of the balanced line. Enter the first and last object of the balanced line into the cycle object. The following example demonstrates how the cycle object works.

Example 17: Cycle

Three machines will be synchronized. Create the following Frame:

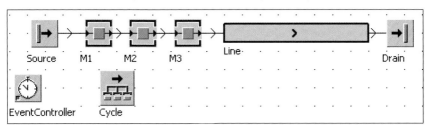

*Settings: Source: interval 2 minutes; M1: processing time 2 minutes; M2: processing time 2 minutes; M3: processing time 1 minute; Line: length 8 meters, 8 minutes time. Let the simulation run. The station M3 is finished earlier than the other stations and then empty for a while. Now double-click the cycle object. Enable the cycle object with **ACTIVE** and enter the first and the last station:*

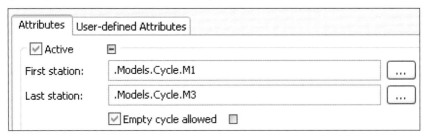

The part on the object M3 remains on the station until all other stations are finished too.

3.2.9   The Store

The store has an unlimited number of storage places, which are organized in a matrix. As long as one place is free, the store can receive MUs. Without a control method, the store places the MU on any free place in the matrix. As opposed to the active material flow objects, the store has no setup time or processing time and no exit controls. The MUs remain in the store until they will be removed by using a control.

When you want to reduce the dimension of a store, you can only eliminate empty places. If MUs are still located on the places you want to delete, then you have to first delete the MUs or transfer them to other places within the smaller dimension. If the store is failed, it cannot receive MUs, but it can move MUs out of it.

3.2.10 The Line

3.2.10.1 Behavior of the Line

The Line is an active material flow object. It transports MUs along a route with a constant speed (accumulating conveyor like gravity-roller conveyor, chain conveyor). MUs cannot pass each other on the line. Unless you have entered an output control or you have chosen a different behavior, the Line distributes MUs to its successors. When an MU cannot exit (e.g., occupation of the Successors), the setting "Accumulating" determines whether the MUs maintain their distance or move up.

3.2.10.2 Attributes of the Line

Attributes	Times	Failures	Controls	Exit Strategy	Statistics	Curve	Custom A ◀ ▶

Length:	2	⊟ m	☑ Accumulating		☐
Speed:	1	☐ m/s	☐ Backwards		☐
☐ Acceleration		☐	Acceleration:	1	☐ m/s²
			Deceleration:	1	☐ m/s²
Time:	0:02				
Capacity:	-1	☐			
MU distance:	-1	☐			

Length: Length of the line (the maximum number of MUs on the line is calculated by dividing the length of the line by the length of the MUs).

Speed: The line has the same speed along the entire length. You can set the speed to zero to stop the line.

Time: Enter the time, a MU needed for transportation from the beginning until the end of the line (the speed is calculated thereof).

Capacity: The capacity determines the maximum number of MUs, which can be positioned entirely or in part on the Line (-1 for an unlimited capacity).

Accumulating: See the following example.

Example 18: Line 1

Create the following Frame: Source: each 6 seconds one part, Line: length 18 m, 1 m/s speed, Drain: processing time 0 seconds

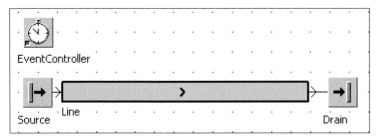

The default setting of the Line is ACCUMULATING (a checkmark in the box).

Now fail the drain (checkbox FAILED) and save your changes. Start the simulation. The MUs move up. This way, the Line works like a buffer. The simulation only stops when the entire line is occupied with MUs:

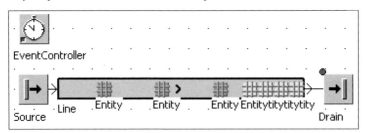

Remove the MUs from the simulation model (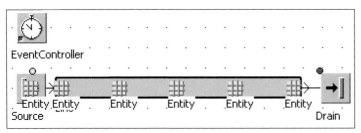*), then clear the checkbox ACCUMULATING in the dialog of the Line and confirm your changes by clicking OK. Now restart the simulation. The parts on the line keep their distance. With this setting, the line cannot be used as a buffer.*

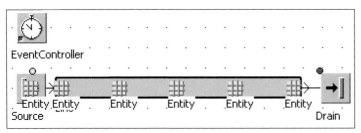

What kind of behavior the line must have depends on the technical realization of the Line. Conveyor belts or roller conveyors are normally accumulating; a chain conveyor is generally non-accumulating.

Backwards: The line can move forward or backward. If it should move forward, the checkbox **BACKWARDS** is cleared.

Example 19: Line 2

Work stations are arranged around a conveyor belt.

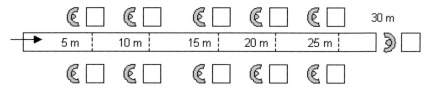

One part arrives every 5 seconds, the processing time of the APs is 10 seconds, and the last job requires 5 seconds. The speed of the conveyor is 1 m/s.
  Create the following Frame:

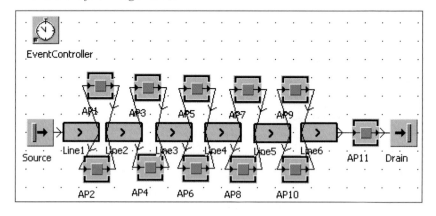

A Side Note to Inheritance

If you want to increase the speed of the entire conveyor belt, you have to alter the speed of each line segment. The same applies, e.g., for the work places. For this problem, there is a simple solution. Define an appropriate object in the class library and change its properties there. Create derivatives by dragging the object from the class library into the Frame. Derivation creates Children (Child objects). The main object is called base object/base class (Parent/parent class). The child objects inherit all properties and methods from its base class. There is also a link between basic class and child objects (inheritance). If you change the properties of the base object in the class library, you are also changing the properties in the child objects (if inheritance is switched on).

3.2.10.3 Curves and Corners

Lines may have a very complex course. Plant Simulation allows you design the course as complex as it is in the real layout. If you have inserted a line in a Frame, you can extend it by dragging. You can change the shape of the Line with the context menu command **APPEND POINTS**. Plant Simulation draws the length of the line by a setting ratio of meters to pixels. In the basic setting, the grid spacing of a Frame is 20 x 20 pixels. You can define a different grid space in Plant Simulation window under **TOOLS – PREFERENCES – MODELING**.

You can adjust the ratio of grid and dimensions for each Frame individually. Select **TOOLS – SCALING FACTOR ...** in the Frame window.

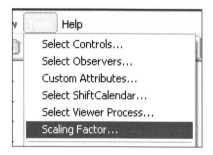

Enter the required size ratio into the following dialog:

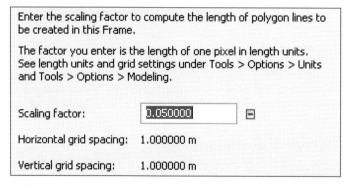

Define the visual appearance of the curved object on the tab **CURVE**

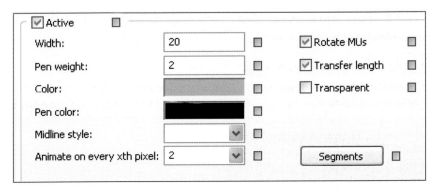

Clear the checkbox **ACTIVE** if you want to use a separate icon for the line (e.g., from an icon library). If you append points (right mouse button – **APPEND POINTS**) and hold Ctrl + Shift, you can draw arc segments:

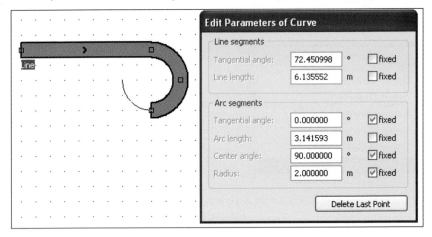

3.2.11 AngularConverter and Turntable

AngularConverter and Turntable assist you in modeling curves or junctions on Lines. Oftentimes independent technical solutions are required, which themselves require a certain amount of time for implementing them. Plant Simulation offers three options:

1. Append a corner point to the line, and extend the line in a 90-degree angle. Without SimTalk you cannot implement a special (higher) time for the transfer.
2. You can use the object AngularConverter. You can simulate processes in which the part is transported to a certain point, stops, and then accelerates again at an angle of 90 degrees. Retarding and acceleration times will be considered as time (e.g., 4 seconds). The entity will not be rotated during this process.

3. If the part is to be rotated while transferring (via a robot or a turntable, for example), you can use the object TurnTable.

Example 20: Turntable and AngularConverter

For comparing the various solutions, here is a small example: Duplicate an entity, and rename it to partarrow. Change the icon so that it matches the following picture:

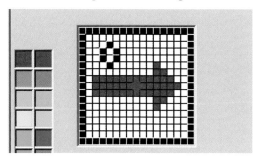

Create the following Frame:

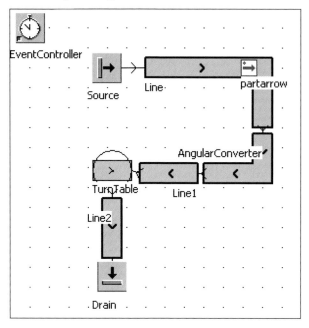

Enter the following settings: The source generates the MU partarrow at intervals of 1 minute. Leave the basic settings for all other objects. Start the simulation and track the movements of the part.

3.2.11.1 Settings of the AngularConverter

You can select different lengths and the associated speeds:

Entry length:	2	▣ m
Exit length:	3	▣ m
Entry speed:	1	☐ m/s
Exit speed:	1	☐ m/s

The **MOVING TIME** (Tab Times) is the time which the converter needs to switch from one direction to the other.

| Moving time: | Const ⌄ | 0:03 | ☐ |

3.2.11.2 Settings of the Turntable

The Turntable accepts a part and rotates it by 90 degrees and from there moves it on in the direction of the connector. If you select "**GO TO DEFAULT POSITION**" (and possibly enter an angle), the turntable returns to this position after moving the part on. If the option is not selected, the turntable only rotates when the next part is ready to be moved.

Length:	2	▣ m	Entry Angle Table	▣
Rotation point:	1	☐ m	Exit Angle Table	▣
Conveyor speed:	1	☐ m/s		
Rotation time per 90 °:	0:04	☐		
Rotate when:	Centered ⌄	▣		
☑ Go to default position	▣	Default angle:	0	☐ °

3.2.12 The PickAndPlace Robot

3.2.12.1 Basic Behavior

From version 9 on Plant Simulation provides the object PickAndPlace. You can easily model robots with it, which pick up parts at one position and rotate and place the parts at another position. Plant Simulation determines the necessary angles of rotation according to the position of the successors or you can enter the angles into a table.

Example 21: PickAndPlace 1

Create the following Frame:

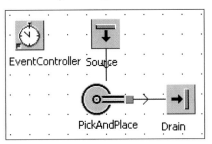

The source produces one part every minute. All other components use their basic settings. When you start the simulation, the pick and place robot transports parts from the source to the drain.

3.2.12.2 Attributes

If you connect the pick and place robot with other objects with Connectors, Plant Simulation generates a table of positions and associated objects. You can find the table in the tab **ATTRIBUTES**. Click the button **ANGLES TABLE**:

The position of 0° corresponds to the so-called 3 o'clock position. The angles are specified clockwise.

	Name	Angle
1	Source	270.00
2	Drain	0.00

The **TIMES** table controls the time consumption between the different rotation positions. Here you define the duration of the movement from one position to another. Default is one second. To change the times, click the button **TIMES TABLE** on the tab **ATTRIBUTES**:

The default angle designates a waiting position, which the PickAndPlace robot takes when you select the respecitve option on the tab **ATTRIBUTES**.

0.0000			
Full/Empty	Default Angle	Source	Drain
Default Angle	0.0000	1.0000	0.0000
Source	0.0000	0.0000	1.0000
Drain	1.0000	1.0000	0.0000

Enter the duration of the rotations like this:

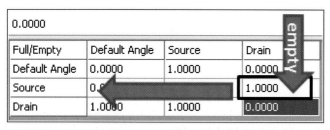

GO TO DEFAULT POSITION

In its basic setting, the PickAndPlace robot waits at the unloading position, until a new part is available at the loading position. Then it turns to the loading position and loads the part. If you select "**GO TO STANDARD POSITION**" the robot then moves to this position after placing the part.

| ☑ Go to default position | ⊟ | Default angle: | 270 | ⊟ ° |

Example 22: PickAndPlace 2

A Pick-and-place robot is to sort parts. Red, green, and blue parts arrive in a mixed order. The robot is to distribute those according to the value of the attribute "col" of the parts. Create 3 parts (part1, part2, part3) in the class library. Assign a user-defined attribute ("col", data type: string) to all 3 parts. Set the values to "red", "green", and "blue". Color the parts accordingly. Create the following Frame:

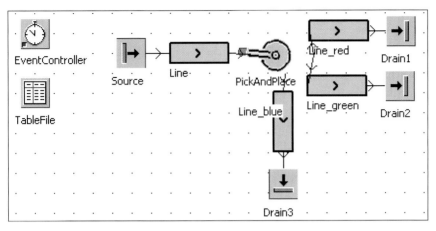

*Settings: The source randomly creates part1, part2, and part3 33% each in an interval of 2 seconds. Connect the PickAndPlace object with Line_red first, then with Line_green, and lastly with Line_blue. You can define the distribution of parts by color by the exit strategy of the PickAndPlace robot. Select the option **MU ATTRIBUTE** on the Tab **EXIT STRATEGY**. Then click Apply. The dialog shows additional dialog items:*

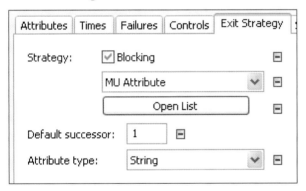

*Select the **ATTRIBUTE TYPE**: String (the data type of the user-defined attribute col). Click **OPEN LIST**. Enter the attributes, the values and the successors which should be transferred here.*

	Attribute	Value	Successor
1	col	blue	3
2	col	green	2
3	col	red	1

Finally, the robot should place the parts assorted by color to the lines.

3.2.13 The Track

The track is a passive object for modeling transport routes. The transporter is the only moving object, which can use the track. The dwell time on the track is calculated using the length of the track and the speed of the transporter. MUs cannot pass each other on the track (retain their entrance order—FIFO). If several transporters with different speeds are driving on the track (a faster one catches up with a slower one), a collision occurs. The faster transporter automatically adjusts its speed to the slower one. With a capacity of -1 the maximum capacity of the track is determined by the length of the track and the length of the transporters (length of 10 m and 1 m per transporter result in a maximum of 10 transporters); otherwise, the capacity is limited by what you enter.

Attributes:

Length:	2	▣ m
Capacity:	-1	▣
Backward destination list:		... ▣
Forward destination list:		... ▣

Plant Simulation determines the length of the track if you activated the checkboxes **ACTIVE** and **TRANSFER LENGTH** on the tab curve.

   Backward/Forward destination list: A track can connect several workstations. You can specify which stations have to be covered on the route (forward and backward).

Example 23: Track

Create the following Frame:

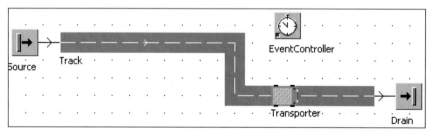

You have to "convince" the source to produce transporters. Open the source by double-clicking it. Enter a time of 1 min as the interval for the production. Select.MUs.Transporter as the MU.

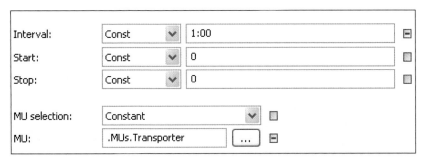

When you start the simulation, the source produces a new transporter every minute and moves it onto the track. The transporter moves with the speed you entered on the track and will be moved to the drain at the end of the track. The drain destroys the transporter.

Rotate MUs
Clear the option **ROTATE MUS** on the tab Curve, and test what happens:

The option **ROTATE MUS** animates the transporter, so that it always points forward (the front always points in the direction of movement). Therefore, the icon of the transporter rotates. Try it once with a curve (like the Line: Context menu, then attach corner points, and insert the curve with CTRL + SHIFT).

3.2.14 The Sorter

3.2.14.1 Basic Behavior

The sorter can receive a certain number of MUs and move them on in a different order. The removal order of MUs, which the sorter contains, depends on definable priorities. The MU with the highest priority will be transferred first, no matter when it entered.

The following selection criteria are offered:

- Duration of stay
- MU-attribute
- control

The content of the sorter is sorted, if either

- a MU enters the sorter or
- the content of the sorter is accessed.

When several MUs have the same value within a sort criterion, then the order of these MUs remains undefined. You can use the sorter for simulating queue logics.

3.2.14.2 Attributes of the Sorter

A number of rules (e.g., shop floor management) exist for controlling queues. An important criterion is, for example, the throughput time of an order (from entry into the production to delivery to the customer). Special orders from major customers are often preferred to be able to deliver quicker. The throughput time of the remaining orders thus increases. Another rule is, for example, that the order which causes the least retooling cost will be fulfilled first. The simplest queue management follows the first come, first serve principle (or first in, first out)…

Capacity:	100
Order:	Ascending
Time of Sort:	On Entry
Sort criterion:	Occupation Time

Start Sorting

Capacity: Enter the number of stations in your sorter. "-1" stands for an unlimited capacity. You can access individual stations by their index ([…]).

Order: The sort order determines whether the MUs are sorted in ascending or in descending order. (Priority 1 very high; capital commitment, …).

Time of sort: When should be sorted? If **ON ENTRY** is selected, newly entering MUs will be sorted into the existing order of the other MUs. The sequence is not updated, even if the values of the sort criteria change. When you selected the option **ON ACCESS**, the MUs will be sorted dynamically. At each entrance of a new MU or before moving it, the MUs will be reordered (taking into account the current values of the sort criterion).

Sort criterion: Sort criteria can be:

- Occupation Time: The MUs will be sorted according to their occupation time in the sorter (descending: first in-first out or ascending: first in-last out)
- MU-Property: You can enter order attributes and statistical values (statistical values only if statistics for the MUs is active).

Example 24: Sorter

A process with an availability of 50% will be simulated. A sufficiently large buffer is located in front of the process. The parts will be processed after blockages with different priorities. We want to achieve parts with higher priorities that have a much lower throughput time than the parts with lower priority. Create a folder color_sorting below models. Create a Frame within the folder color_sorting (Right-click the folder icon – NEW – FRAME). Duplicate all the required objects in this folder.

Frame:

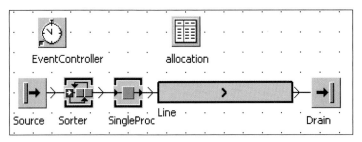

Settings:

1. *Insert three entities, and name them red, green, and blue. Assign them different colors (recommended: 5x5 pixels, colors according to the names) to better distinguish them. Open the source, and set the following values: Interval: constant 2 seconds, MU-Selection: random, table: allocation, enter the following values into the table allocation (Note: You can insert the addresses using drag and drop; Drag the relevant parts from the class library to the table, and drop them there):*

	object 1	real 2	string 3
string	MU	Frequencies	Name
1	.Models.color_sorting.red	0.10	
2	.Models.color_sorting.green	0.60	
3	.Models.color_sorting.blue	0.30	

SingleProc: Processing time: 1 second, Availability: 50%, 30 minutes MTTR (based on the simulation time), Line: length 8 meters, 1 m/s speed, accumulating, Drain: 0 seconds processing time

2. *User-defined Attributes*
Create a user-defined attribute for each part (Double-click the part in the class library, tab CUSTOM ATTRIBUTES):

Custom Attributes

Name: importance

| Value | Statistics | Communication |

Data type: integer

Value: 1

Set the following values for the attribute "importance": red: 1, blue: 2, green: 3. The parts in the sorter should be sorted according to the attribute "importance".

3. Sorter attributes
In the sorter, you have to select the criteria by which it sorts (default on entrance of a new part into the sorter).

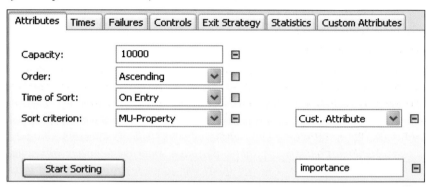

Ascending order: The part with the smallest importance "scrambles forward"; it will be sorted in ascending order, Sort criterion: Cust. Attribute – importance. Click Start Sorting to start the sorting process.

4. Control results
*Run the simulation for a while. If a failure takes place, you will see a series of red parts exiting the SingleProc (the parts enter the sorter in a mixed order). Look at the type-specific statistics of the drain. Click the button **DETAILED STATISTICS TABLE** on the tab **TYPE SPECIFIC STATISTICS**:*

The value LT_Mean shows the average throughput time of the parts. The part red has a much lower throughput time than the part green:

	Type	Time	Total throughput	%Parts	LT_Mean	L
1	blue	2:07:45:16.0	30240	30.49	13:21.3454	2
2	green	2:07:45:16.0	58996	59.48	1:01:45.7411	2
3	red	2:07:45:16.0	9958	10.04	8:57.8818	1

3.2.15 The FlowControl

3.2.15.1 Basic Behavior

The FlowControl itself does not process MUs. The Flow Control is always positioned between two or more other objects and defines the flow behavior between these objects. If needed, you can also combine several FlowControl objects.

3.2.15.2 Attributes

Example 25: FlowControl 1

You are to take a scrap rate at a workplace into account. Scrap represents a branch in the flow of materials (e.g., the quality control sorts a defective part, the defective part is moved to the drain). To simplify the presentation of scrap, the parts have the property "io" with a value true or false in the simulation. You can branch the flow according to these values.

Create the following Frame:

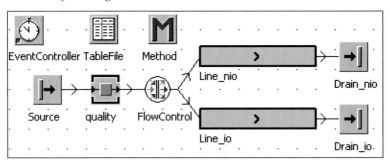

*quality (quality assurance): 1 minute processing time, Lines: 3 meters, speed 1 m/s. The order in which you insert the connectors determines the number of the successors. First, connect the FlowControl with the line_io, then the the Flow-Control with line_nio. Duplicate an entity, and name it "part". Create a user-defined attribute for the part: Name: io, Data type: boolean. 10% of the parts should have the value "false" for the attribute "io". The remaining parts will receive the value "true". The allocation will be done randomly. For the assignment, you can use the source. In the dialog of the source, select the option **MU-SELECTION – RANDOM**. Select the TableFile for the allocation.*

Open the TableFile by double-clicking it. Enter the following values into the table:

	object 1	real 2	string 3	table 4
string	MU	Frequencies	Name	Attributes
1	.MUs.part	0.90		attribute
2	.MUs.part	0.10		attribute

*Enter the same part twice. Enter a name in the column ATTRIBUTES (in the above case "attribute" is used as an internal name); Plant Simulation then creates a table. Press the F2 key in the field **ATTRIBUTES**. It opens another window. Enter the name of the attribute (io) and the value (true/false) into the field with the correct data type (boolean):*

	string 1	integer 2	boolean 3	string 4
string	Name of Attribute			
1	io		true	

Proceed as described with the second row for the part (boolean false). You can now evaluate the attribute io in the flow control. Select Method on the tab Exit Strategy and select the method:

Exit Strategy	Entry Strategy	Custom Attributes

Strategy: Method   ▼ ▬   ☑ Blocking

Method: Method   [...] ▬

Within the specified method, you can access the respective part with @; for the successor to be moved, an integer has to be returned (1 or 2 in this example). Enter the following source code into the method:

```
(r : integer) : integer
is
do
   if @.io=true then
      return 1;
   else
      return 2;
   end;
end;
```

Run the simulation and check the results using the type-specific statistics of the drain.

Attributes of the FlowControl

Tab **EXIT STRATEGY**: Here you set the exit behavior of the FlowControl. **BLOCKING** means, that if the successor cannot receive parts, the FlowControl waits until it can receive parts again.

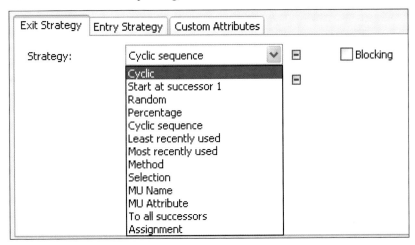

Here is a short selection of the strategies:

- START AT SUCCESSOR 1: The flow control attempts to always pass the MU on the successor number 1. If the successor 1 is always receptive, each MU will move to it. The MU will be passed onto the next successor only if moving is not possible (faulty, occupied).
- CYCLIC: The FlowControl tries to move the MU based on the recent passing on the next object (in the list of successors).
- SELECTION: The FlowControl tries to move the MUs onto the successor that meets a certain property.

You can also use a method for distributing the MUs: Specify a method, which returns the number of the successor. You can access the MU, which is to be transferred with @. If the part cannot be moved to the designated successor, the method will be called again.

Example 26: FlowControl 2

MUs should be separated by their name. Three parts (part1, part2, and part3) will be stored together on the same place (buffer1). After this, the different parts will be processed on separate machines.

Create the following Frame:

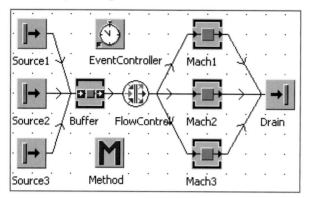

Data: Buffer: capacity 100 parts, Mach1: 2 minutes processing time, Source1 (part1): interval 6 minutes, Source2 (part2): interval 6 minutes, Source3 (part3): interval 6 minutes, Mach1, Mach2, Mach3: each 6 minutes processing time. Failures will not be considered. Select Method as exit strategy in the dialog of the flow control and select the method in your Frame. You can use "@" for accessing the current MU and you have to indicate the number of the successor. The source code in the example above might look as follows:

```
(r : integer) : integer
is
do
    if @.name="part1" then
            return 1;
    elseif @.name="part2" then
            return 2;
    elseif @.name="part3" then
            return 3;
    end;
end;
```

Percentage: You can select a percentage distribution. The basis of this is a distribution table. In this table you enter the percentage for each successor.

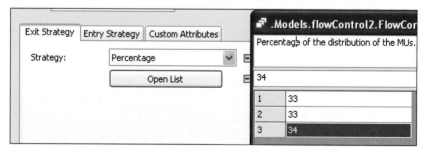

Random: You can define a distribution function for the transfer. The distribution function is used to determine the successor.

Cyclic sequence: If you select Cyclic sequence, then the MUs will be passed in a defined sequence to the successors. The order is entered in the corresponding table.

To all Successors: This distribution creates duplicates of MUs. Any successor will receive a duplicate (always blocking).

Assignment: There is only one successor. You can set a property of the MU using a method …

MU Attribute: Here is the successor chosen by the value of an attribute of the MUs.

Tab Entrance Strategy

Under this tab you set the reunification strategy of the FlowControl (several pre-decessors).

Example 27: Just in Sequence

With the FlowControl, you can model just in sequence processes. In such processes, parts from different sources are used for production and assembly processes in a specific order (sequence) and the parts will be transported in this order to the workstations. In the following example, three different sources deliver parts. These parts must be placed in a sequence (two parts of Source_r, one part of Source_g, two parts of Source_b). They will be transported in this order to an assembly station, which places five parts each on a palette. Create the following Frame:

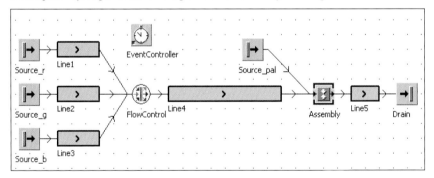

Settings: Create 3 entities (red, blue, green; each 5x5 pixels, color the parts). Source_r produces one part "red" every 4 seconds, Source_g one part "green" every 8 seconds, and Source_b produces one part "blue" every 4 seconds. The parts need 10 seconds from the entrance to the exit of Line1, Line2, and Line3. The travel time on the Line4 is 20 seconds. Source_pal produces a container (ca-pacity 5 parts) every 8 seconds. Set the assembly station so that 5 parts of Line4 are loaded onto the container. The assembly time is 8 seconds. Failures and breaks are not taken into account. Click the tab ENTRANCE STRATEGY in the dia-log of the flow control. Select the strategy CYCLIC SEQUENCE.

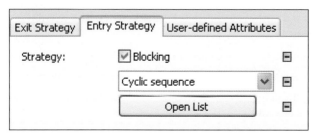

*Clear the inheritance button next to the button **OPEN LIST** (click the green button). Click the button **OPEN LIST**. Enter the order of the "predecessors" into the list. For the example above, the sequence is 1-1-2-3-3:*

.Models.justInSequence.FlowC	
Sequence of access attempts.	
1	
1	1
2	1
3	2
4	3
5	3

*So that the sequence will remain intact even if a part is temporarily not available, select **BLOCKING**.*

3.3   Resource Objects

3.3.1   Usage and Example

Resource objects can be used for the simulating employees. The simulation of employees is especially interesting for these constellations:

- Repairs
- Machine operation (the machine cannot work without operators)
- Employee as transporter of parts (carries parts)

Example 28: Resources for Repairs

The same employee maintains two machines. Both machines have a processing time of 2 minutes, an availability of 80%, and a MTTR of 45 minutes. The staffroom of the service employee is 50 meters away from the machines. Create the following Frame:

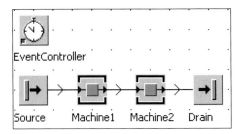

3.3.2 The Worker-WorkerPool-Workplace-FootPath Concept

Workers are generated in the worker pool and stay there. The workers offer various services (e.g., repair, operate, and tool change). A broker mediates the workers to the individual workstations when they request the services. The broker sends the workers from the worker pool to the machines. If there is a footpath, then the worker walks from the worker pool to the machines on the footpath. The worker stays on his workplace while doing his job.

To simulate the workers, you need the following objects:

1. Broker
2. WorkerPool
3. Worker
4. FoothPath
5. Workplace

Complete the following example. It is best to begin with the broker. Then drag the worker pool into the Frame so that it is placed close to the broker. Next, drag the workplaces into the Frame. Drop it close to the machine. Then insert the footpaths into the Frame. You can insert the footpaths with corner points, just like tracks and lines. Connect the WorkerPool with the footpath and the footpath with the work places:

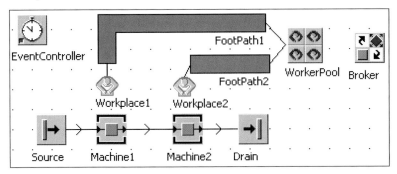

3.3.3 The Broker

The broker mediates between suppliers and demanders of services.

You do not have to select settings in the broker. You have to propagate the machines as well as the resource pool to the broker.

3.3.4   The WorkerPool

The worker pool produces a number of workers according to a creation table and makes them available for the registered services after a request. If the workers do not work, Plant Simulation shows them on the worker pool.

If you insert a worker pool close to a broker, the broker will be entered into the field **BROKER** you have to enter the workers, which are managed in the pool, into the creation table. Click on **CREATION TABLE**:

	Worker	Amount	Shift	Speed	Effic	Additional Services
1	*.Resources.Worker	1				repair

Drag the workers to the creation table, and enter the amount. In the column **ADDITIONAL SERVICES,** you can enter designations for services. The services are marked by a string. So you can enter any of your needed services. If you want to register more than one service, then you can do it in the columns to the right next to **ADDITIONAL SERVICES** (maximum 30).

Example: Enter for the Example the service "repair".

You can also enter services directly in the dialog of the worker.

Get job orders in the pool only: If this option is selected, the worker must return to the worker pool between the individual orders (walk). If this option is cleared, then the worker can also receive an order en route to a job.

Workers can beam to the workplace: If this option is selected, the worker can walk directly to the workstation, even if there is no suitable footpath. If this option is cleared and if there is no footpath to the workplace, Plant Simulation outputs an error.

Workers can work remotely: If this option is selected, the worker can do his job, even if the respective workplace is already occupied (e.g., by another worker). Otherwise, you will get an error.

Broker: Select the broker who will broker between the work places and the worker pool (you can drag the broker to the respective field of the worker pool).

3.3.5   The Worker

You have to set Worker-related settings in the class library because the instantiation of the worker is realized by the worker pool.

The worker has a number of properties, which are important for the simulation:

Priority:	0	⊟	Supply:	Services	⊟
Efficiency:	100	⊟			
Speed:	1.67	⊟ m/s			
Capacity:	1	⊟			
Shift:		⊟			
Broker:		... ⊟			

Priority (between 0 and 10): The higher is the priority, the sooner a job will be performed.

Efficiency (in percent): 50% means that the worker needed twice the time for the job.

Speed: Speed of the worker on the footpath.

Capacity: Number of parts, which the worker can carry at once.

Shift: Name of the shift, during which the worker works. If no shift is entered, the worker can work in all shifts.

Broker: The broker will be assigned by the worker pool.

Click the button SERVICES to assign a range of services to the worker:

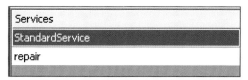

Services
StandardService
repair

3.3.6   The Footpath

On the footpath, the workers move between work places and the worker pool. For that, they consume a time, resulting from the speed of the worker and the length of the footpath.

3.3.7   The Workplace

The worker stays on the workplace when he performs a job. Only one worker can stay on a workplace at any one time.

With the workplace, you connect an event of the machine (e.g., SingleProc) with a request for a service (e.g., failure: request of the service "repair"). The workplace needs to be assigned to another object, which happens automatically when you place the workplace close to another object. You can also select the respective object in the dialog:

Enter "repair" into the list Service:

The machine itself must request the service. You need to designate the broker and determine the service which will be required. Do this in the dialogs of the objects. The service will be requested if the machine fails. The failure importer is responsible for the request of the relevant services (tab Failure Importer):

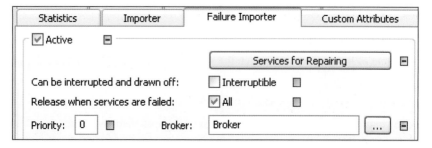

Activate the Failure Importer. Turn off inheritance for the services list (Click on the green box next to the button Services for Repairing). Click the button "Services for Repairing". Enter the service name (repair) and the number of workers.

Service	Amount	A
repair	1	

Select the broker.

Confirm your changes and run the simulation for a while. If a failure occurs, the worker moves from the workerpool to the machine and remains there until the failure is removed. Then he moves back into the workerpool.

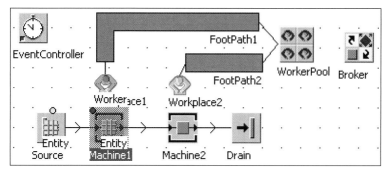

3.3.8 Worker Transporting Parts

You can also use the worker for transporting parts. That is important, for example, if you need to simulate a multi-machine operation, in which workers have to transport the parts from one machine to the next. The active material flow objects SingleProc, ParallelProc, Buffer, PlaceBuffer, Sorter, and Source have the exit strategy: Carry Part Away.

Example 29: Resources Exit Strategy Carry Part Away

We want to simulate a multi-machine operation. A worker carries parts from a buffer and carries them to a Machine1. If the Machine1 has finished, the worker carries the parts from Machine1 to Machine2 and then to another buffer.

Create the following Frame:

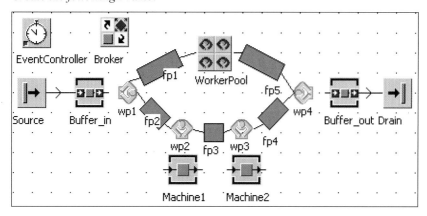

Settings: Source: Interval: 1:30 minutes, Machine1, Machine2: 0:50 minutes processing time, Buffer: capacity four parts each, 0:30 processing time, worker: one worker, service: "carry part away", broker and workerpool, assign the work

places to the associated machines. The worker transports the part from one job to the next. Therefore, the next workplace must be assigned to each machine and each buffer. Open the dialog of Buffer_in. Click the tab exit strategy, and select Carry Part Away from the drop-down list. Then click Apply:

Select the broker. Specify the workplace, which next handles the part. Enter a maximum dwell time which determines how long the worker waits for a part from a machine (or the time, which is necessary for unloading/loading of the part). If you enter a longer dwell-time than the processing time, the worker stays at the machine during processing and carries the finished part to the next station.

Do the same for Machine1 and Machine2. The worker now carries the parts from one machine to another. The worker does not check whether the succeeding machine can receive the part. Therefore, the worker might carry a part to an already occupied machine.

3.4   General Objects

3.4.1   The Frame

3.4.1.1 General

The Frame is the base of all models. You can use Frames to create your own objects with an arbitrary behavior.

You can combine various basic objects to an object with higher functionality. The new object (Frame) can be used like any other object. The interface between different user-defined objects (Frames) is the object interface. Thus, it is possible to build a model in several hierarchies. The object Frame does not have its own basic behavior. You can create new Frames on the context menu of the folders of the class library. Click the right mouse button on a folder icon and select New – Frame.

3.4.1.2 The Frame Window

 Opens the Frame window in which the object is located.

 Opens the Frame from which the current Frame was derived.

 Selects all the objects that have unconnected entrances or exits. Already selected objects remain selected.

 Opens the icon editor.

 Deletes one or more objects after confirmation.

 Deletes all mobile objects on all objects.

 Shows or hides object names.

 Shows or hides connectors.

 Shows or hides comments.

 Shows or hides the grid Frame.

 Activates or deactivates the menu command CHANGING STRUCTURE.

 Starts the simulation

 Resets the simulation

 Starts the simulation without MU-animation

 Opens the toolbox for creating graphics in the Frame.

 Shows the Eventcontroller. If no Eventcontroller is inserted in the Frame, it will be inserted in the Frame.

The menu Icons is important for working with the Frame:

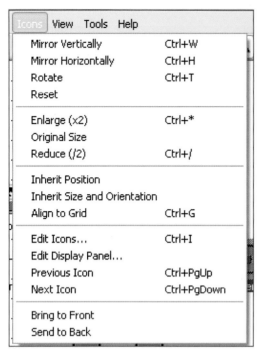

Here, you can rotate, shrink, and mirror the icons of the objects.

3.4.2   The Connector

3.4.2.1 Basic Behavior

The Connector connects objects and Frames. Clicking the Connector in the class library or the toolbox activates connection mode (Mouse pointer changes to the icon Connector). To connect object A with object B, first click on object A and then on object B (the connection will only appear if the menu command View-Options Connections is enabled). An arrow shows the direction of the connection:

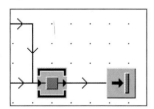

You can also insert corner points. You can create a rectangular corner point by holding down Shift + clicking the left mouse button:

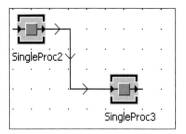

Follow these steps (Connection mode): Click the left mouse button and hold down the Shift key on the location of the corner point, then click the position of the second corner point, and then click the icon of the other object.

3.4.2.2 Attributes

You can view the properties of the connector by double-clicking the connector in the Frame (for this special connector) or in the class library for all connectors by double-clicking the icon of the connector. You can enter the width of the connector in pixels (allowed: 1–100) and you can select a color.

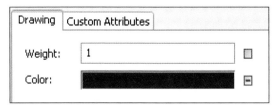

The visibility of the connectors is generally controlled in the Frame window:

3.4.3   The EventController

3.4.3.1 Basic Behavior

The EventController coordinates the events during the simulation.

When an MU moves onto a SingleProc, the EventController calculates when it will be passed through the SingleProc and will exit. This time 10 seconds, for example, is entered in a table of the EventController. The EventController then informs the SingleProc that it has to process an exit event. The MU will be moved to the succeeding object.

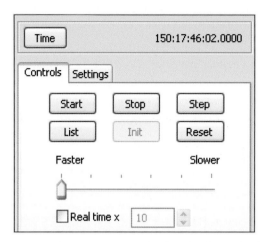

Time: There are two modes for the time: absolute time and date. When you start the simulation, the simulation time is set to zero and then starts. For the date format, click the button Time:

The simulation will start at a specified date. You can set the date to the required value in the tab settings:

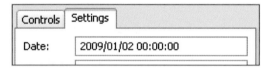

Clicking the button TIME toggles between the modes.

Tab Controls:

Start: Clicking the Start button starts the simulation. If you previously clicked Reset or it is the first simulation run, the INIT control in the model will be called. You can also start the simulation by pressing the Shift key and double-clicking the EventController in the Frame.

Stop: Clicking Stop stops the simulation after processing the current event. You can also double-click the Eventcontroller while holding down the Shift key.

Step: With Step you can run your simulation step by step. It processes only one event.

Reset: Reset calls all Reset controls of the model. All unprocessed events will be deleted and the simulation time will be set to zero. Statistics will also be reset. Any failure will be removed (Pauses are not changed).

List: Clicking List opens the window of the event debugger. Here, all events are listed in ascending order.

Init: Init calls all Init methods of the Frame. When you first start the simulation, Init will be called automatically.

Real Time x: The simulation usually jumps from event to event, without waiting for the intermediate times. Real time forces the Eventcontroller to run the simulation during this time as well, including graphic animation, even if no event occurs (you can determine how fast the simulation runs).

Tab settings

Date: Start date of the simulation.

End: Here, you can enter a relative time for the length of the simulation run. This time will be compared with the current simulation time. If both values are equal, the simulation is aborted.

Statistics: The date entered here will reset statistics. This setting, you can use to hide the filling with parts of your model (ramp up) from the statistics.

Backwards: The absolute time is running backwards.

Delete MUs on reset: All MUs in your Frame will be erased before the next simulation run. These also include containers and transporters.

Step over animation events: With this option, events will be processed, after pressing step, as long as a new animation event occurs (graphic moves).

3.4.4 The Interface

3.4.4.1 Basic Behavior

The Interface object allows connecting different Frames.

Example 30: Frame and Interface

Preliminary remark: A basic concern of many object-oriented techniques is reusability. For this purpose, we use a versatile range of classes (data types). With these small building blocks (the same for all projects), we can build the appropriate solutions. For scheduling, a multitude of such "kits" exists, for example "methods time measurement". Using this methodology, human work is broken down into elementary movement objects. These movements will be assigned a time depending on various factors (time studies). A simplified description of the Methods Time Measurement procedure follows. It breaks down human labor into 15 basic tasks (here objects).

Deburring: A worker takes a part off the line, he clamps it, he deburrs it, and he then puts it back on the line. At work, he stands with his back to the line.

The job of the worker can, according to MTM, be broken down into the following task (work begins after loading the finished part onto the line):

Basic objects (abridged)	Duration (sec)
Reach (workpiece)	0.25
Grip (workpiece)	0.1
Bring	0.3
(Body) rotation (to the place)	0.5
Walk	1
Reach	0.2
Handling (clamping)	3
Reach (tool)	0.5
Grip (tool)	0.1
Bring (tool)	0.2
Handling (deburring)	3
Reach (tool)	0.5
Reach (part)	0.3
Grip	0.1
Handling (unfix the part)	2
(Body) rotation (to the line)	0.5
Walk	1
Reach	0.3
Total:	13.85 sec.

Note: To simplify matters, the element "drop" is missing.
You need to create the movement objects as single classes in the class library (file mtm_en.spp):

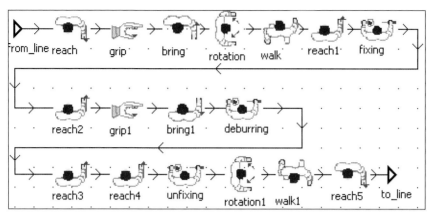

Note: In this example, you have to prevent that more than one part stays in the Frame "deburring". After an MU has entered the object "reach", the entrance of the object "reach" must be locked and may only be opened again if the part "reach5" exits. To do this, you need a method object in the Frame deburring (folder information flow). Add a method object to the Frame. Open the object by double-clicking it. Turn off inheritance (button ⬚) if you cannot enter your source code into the method editor.*
Enter the following source code:

```
is
do
 if ?=reach then
  reach.entranceLocked:=true;
 else
  reach.entranceLocked:=false;
 end;
end;
```

Enter the name of the method into the object reach (tab control→ entrance). Enter the method as exit control into the object reach5, clear the checkmark to the left of front, and select the check box rear.

Times	Set-Up	Failures	Controls		Exit Strate
Entrance:	Method	...	⊟		
Exit:		...	☐	☑ Front ☐	☐ Rear
Set-up:		...	☐		

You can select a new image for the Frame (right mouse button – Edit icons → icon number 1). You can now use the object "deburring" the same way as the other objects in any Frame. Create a new Frame, and drag the Frame deburring from the class library to it:

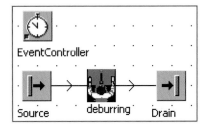

Set the interval of the source to 30 seconds. By double-clicking the object deburring you can open the sub-frame.

3.4.4.2 Attributes of the Interface

Attributes	Exit Strategy	Custom Attributes

Type:		Entrance
Max. number of external connections:	-1	□
Side:	Left ⌄	□
Position in %:	50	□

Type: An interface can be either an entrance or an exit.

Max. number of external connections: Determines how many predecessors or how many successors the device may have.

Position in %: Indicates the position of the connectors (e.g., 0% bottom, 100% top …).

Side: Where the interface on the Frame icon located.

4 Icons

4.1 Basics

It is helpful to use a realistic layout for the simulation model. The layout is divided into several levels:

1. The background of the Frame.
2. The icons of the static objects.
3. The icons of the mobile objects and their movement through the Frame.

All objects of a class share a set of icons. You, therefore, have to duplicate enough classes in the class library for representing a more complex model. For a large quantity of objects, we recommend to organize them in folders.

4.2 The Icon Editor

Every Frame and almost every basic object has a set of icons.

Example 31: Icon Editor

Duplicate a SingleProc in the class library. Open the context menu by clicking the object with the right mouse button and select EDIT ICONS.

This opens the icon editor:

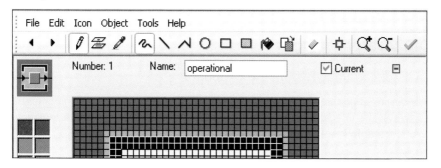

S. Bangsow: Manufacturing Simulation with Plant Simulation, Simtalk, pp. 75 – 83, 2010.
© Springer Berlin Heidelberg 2010

Most objects have two icons:

- No. 0 is the icon that Plant Simulation displays in the toolbar.
- No.1 is the icon that Plant Simulation displays when you insert the object in a Frame.

You can always extend the pool of icons. Select ICON – NEW in the icon editor.

You can replace the existing icon by

1. Drawing a new icon (…). So that the icon is displayed without limitation in the icon library (No. 0); it may have a maximum size of 40 x 40 pixels. All other icons are limited to a size of 4000 x 4000 pixels.
2. Importing existing graphics (file or clipboard).

4.3   Drawing Icons

You can create and edit simple icons in the Plant Simulation icon editor. The following tools are available:

🖉	Freehand line	▣	Filled rectangle
＼	Line	⬥	Fill range
∧	Polyline	▣	Copy range
○	Ellipse	◆	Delete all
□	Rectangle		

You can select colors for the drawing functions by clicking in the color palette or you can select a color from the image with the eye dropper 🖉 tool.

4.4   Inserting Images

4.4.1   Insert Images from the Clipboard

You can insert pictures with copy and paste. Keep in mind that you cannot control the size of the picture after inserting it into Plant Simulation. Copy the image to the clipboard. In Plant Simulation select:

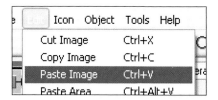

If necessary, the image can be modified in the icon editor. You can insert pictures from some applications in the icon editor with drag and drop. To do so, drag the image from the source application to the icon editor (on the representation of the image).

4.4.2   Inserting Images from a File

You can also create the object icons from existing files.

SELECT FILE – OPEN in the icon editor: …

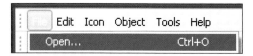

Plant simulation supports importing the following file types:

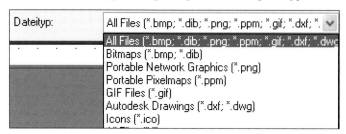

Note that you can even convert CAD drawings into pixel images. When you insert an image from the clipboard or from a file, the transparency information in GIF and PNG files is lost. Plant Simulation provides a transparent color (dark green). It is located at the bottom of the color palette:

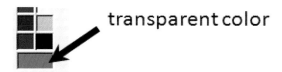

Fill all pixels, which should be transparent in the model with this color. Activate the transparency of these pixels with the command ICON – TRANSPARENT.

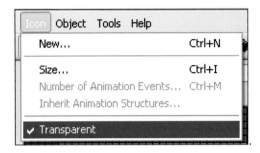

4.5   Changing the Background Color of the Frame

For Frames you can define an image with the name "background". This image is displayed in the background of the Frame. Proceed as follows:
Select **EDIT – NEW** in the icon editor of the Frame:

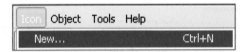

This creates an empty icon. Rename the icon to "background".

Now you can insert a graphic with copy and paste …

Another way is to import an existing file. Select **FILE – OPEN** in the icon editor: The selected file is displayed as the background. If you show everything true to scale, you can generate a realistic model.

Note:
Here, you can also drag a graphic (e.g., an AutoCAD drawing) into a Frame object. A new icon (background) will be created, and the drawing remains visible in the background of the Frame.

4.6   Animation Structures and Reference Points

4.6.1   Basics

When you create your own icons, you have to determine where the MU is displayed on the icon and how the MU (e.g., on a track) will move on the icon. You can set these settings:

- With a reference point on the MU.
- With an animation structure on the material flow objects.

4.6.2 Set Reference Points

The reference point determines where Plant Simulation displays the icon of the part when you insert the object into the Frame or during the animation on another object. If it is an MU, it will be positioned so that its reference point lies on an animation point or an animation line. If the object is a basic object or a Frame, then it will be aligned to the grid in a Frame window. The reference point is positioned on a grid point. The reference point is displayed as a red pixel in the icon editor. You can move the reference point by clicking the **Set Reference Point** button and by clicking a new pixel.

Example 32: Reference Points and Animation Structures

Duplicate the entity in the class library (right mouse button – Duplicate). Name the new class "Part". This part will have a size of 7 x 7 pixels and will be green. Open the icon editor (right mouse button – Edit icons ...). First, change the icon size. Select the menu item: Icon – Size.

Enter a height and a width of 7 and click OK:

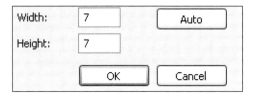

Next, remove everything that is left from the previous icon. Use the "Delete All": ✦ *button. Click on a green color (not the color of transparency in the color palette). First, click the fill icon (🌢) on the toolbar, then click the icon. You have to reset the reference point (in the middle of the icon at the position on 4 x 4 pixels). Click Set Reference Point* ⟊ *on the toolbar. Then select the pixel at the position 4 x 4.*

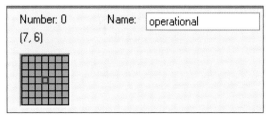

Note:
If you change the size of an MU, you have to do this for the entire set of icons. The entity has two icons that you have to change:

– operational (number 0)
– waiting (number 1)

The fastest way is to copy the current icon (Edit – Copy Image) switch to the next icon and then paste it again (Edit – Paste Image). You have to set the reference point in the second icon (before Plant Simulation version 8).

4.6.3   Animation Structures

In animation mode you can set animation points or animation lines in the icon editor.

Example 33: Animation Structures

Duplicate a SingleProc in the class library. Rename the object to "press". Open the icon editor. Plant Simulation shows a number of simple icons. Select Tools – Clipart Library...

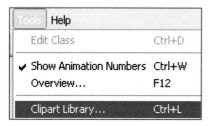

Select the folder machine, and drag the file Press2.gif to the icon editor. Change to animation mode. Click on the icon: . In animation mode, the icon will be shown with a grayed out. The built-in SingleProc already has an animation point. If you change the size of the icon, then you have to manually move the animation point. Click the button . Now you can change the position of the animation point by dragging it. The MU will then be animated on this position on the press.

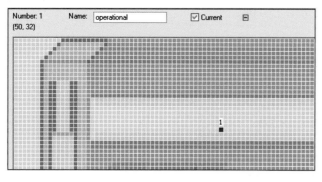

You can also delete animation points and insert new animation points. Place-Buffer, Line, Track, and Sorter use animation lines and animation polylines. If you also want to watch animation, you have to specify how often the MU will be displayed on this line.

Example: Select the PlaceBuffer and open the icon editor. Switch to animation mode. You see a line on the icon. If, for example, more than one animated part is located on the block, you have to specify a number of animation events greater than one. Click on: ICON – NUMBER OF ANIMATION EVENTS …

Enter a number less than or equal to 250 in the dialog that opens. The default is 1, which means that the MU is animated only at the beginning and end of the object, not in between.

4.7   Animating Frames

Often, it is necessary to split simulations that are quite complex. The simulation takes place in various parts or segments, which are connected with each other through connectors. At the top level is an overview of all the objects (Frames) of the simulation. For a better presentation of the sequences, you can animate the flow of mobile units on the Frame.

Example 34: Machine with a Ringloader, Animation on a Frame

We want to simulate a machine. The machine is equipped with a ring loader, which offers 15 places each for unfinished and finished parts. The ring loader is "accumulating". Insert a Frame (e.g., Machine_ringloader). To simulate this machine, we need (for example) three objects:

EntranceBuffer
SingleProc (processing time: 1 minute)
ExitBuffer

Next connectors are necessary for the connections with other objects. The Frame for the machine might look like this:

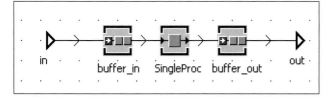

Change the Frame icon (so that indicates the ring loader):

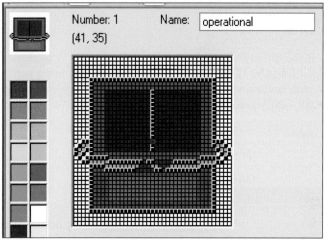

Change to animation mode: Draw three structures on the icon. One each for the:

- *EntranceBuffer*
- *SingleProc (processing time: 1 minute)*
- *ExitBuffer*

You can do this with polylines: 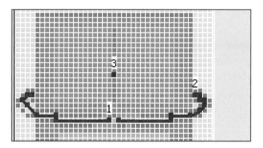 *. Click on the icon Polyline. Then click on separate points; Plant Simulation connects these points with a polyline. To finish the line, click the right mouse button. Plant Simulation numbers your animation structures. It should look like this:*

Now the animation structures on the icon of the Frame must be connected with the animation structures of the objects within the Frame. First, click on the icon

CONNECT ANIMATION POINT... . Then, click on the structure for the input buffer (number 1). It opens a window in which you have to select the respective object and possibly the animation structure of the object. The names of the objects will be listed on the animation structure.

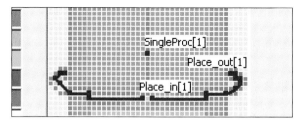

For testing, we build the following Frame:

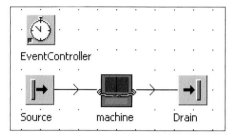

The MUs are now animated on the object (Frame).

5 Programming with SimTalk

The basic behavior of the Plant Simulation objects often in practice is not sufficient to generate realistic system models. For extending of the standard features of the objects, Plant Simulation provides the programming language SimTalk. With it you modify the basic behavior of individual objects. SimTalk can be divided into two parts:

1. Control structures and language constructs (conditions, loops…).
2. Standard methods of the material and information flow objects. They are built-in and they form the basic functionality, which you can use.

You develop SimTalk programs in an instance of the information flow object Method.

5.1 The Object Method

5.1.1 Introductory Example

Example 35: Stock Removal

We want to simulate a small production with a store. The capacity of the store is 100 parts. The workplace produces one part every minute (the source delivers …). Name the Frame "storage".

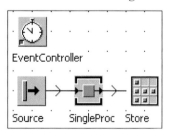

You can create controls with Method objects, which are then called and started from the basic objects using their names.

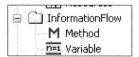

You find the Method in the class library in the folder InformationFlow. Drag a method object to the Frame. A double-click on the icon opens the method.

S. Bangsow: Manufacturing Simulation with Plant Simulation, Simtalk, pp. 85–116, 2010.
© Springer Berlin Heidelberg 2010

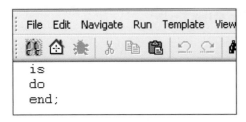

The methods (functions) have always a body:

```
is
do
   -- Statements
end;
```

Declare variables between "is and do", enter your source code between "do and end". First, you have to turn off inheritance. Click on the icon 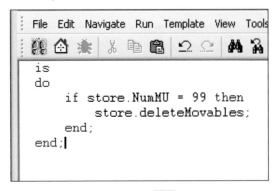 in the editor, you have to formulate the instructions in SimTalk (call your method "stockRemoval"!).

```
is
do
    if store.NumMU = 99 then
        store.deleteMovables;
    end;
end;
```

Confirm your changes with ✓. You now have to assign the method to an object. For this purpose, each object has one or more sensors. When an MU pass through a sensor, the relevant method is triggered. Double-click on STORE – CONTROLS – ENTRANCE; select the correct method, ready we are!

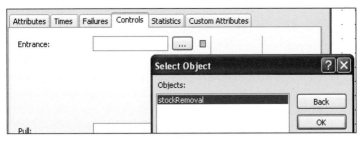

Now start the simulation. If you were successful, then there is no jam. The Store will verify the quantity for each entry. If it is 99, the store will be emptied (a simple solution).

5.2 The Method Editor

Double-clicking a method object opens an editor. You will find a number of functions in the editor, which facilitates your work while programming. If you cannot enter your source code into the method editor, inheritance is still turned on (see above).

5.2.1 Line Numbers, Entering Text

You can display line numbers with the command: VIEW – DISPLAY LINE NUMBERS. The following rules apply for entering text:

- Double-clicking selects a word.
- Clicking three times selects a row.
- Ctrl + A selects everything.
- Copy does not work with the right mouse button (until version 9). Use Ctrl + C to copy and Ctrl + V to insert text or use the menu commands Edit - Copy, etc.
- Use Ctrl + Z to undo the last change (or Edit Undo)…
- Move also works by dragging with the mouse.

5.2.2 Bookmarks

For faster navigation, you can set bookmarks in your code. The bookmarks are displayed in red. To insert a bookmark, select any text and click:

Bookmark functions:

Icon	Description
	Deletes all bookmarks in the method.
	The cursor moves to the previous bookmark.
	The cursor moves to the next bookmark…

5.2.3   Code Completion

The editor supports automatic code completion. If there is only one possibility of completion, Plant Simulation shows the attribute, the method, or variable as a light blue label. You can accept the suggestion with Ctrl + space bar.

```
if store.NumMU = deleteMovables
    store.deleteM;
end·
```

Starting from an object you can display all possible completions. Simply press CTRL + SPACE. In the list you can scroll with the direction buttons, an entry will be accepted with Enter.

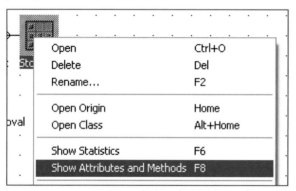

```
do
    if store.N
        store.
    end;
end;
```

addObserver(string,method)
attributeWatchable(string) : boolean
Availability : real

5.2.4   Information About Attributes and Methods

You can always get information about the built-in attributes and methods of an object by Show Attributes and Methods in the context menu of an object.

Open	Ctrl+O
Delete	Del
Rename...	F2
Open Origin	Home
Open Class	Alt+Home
Show Statistics	F6
Show Attributes and Methods	F8

In the table, all methods and attributes are shown (even those you have defined).

Name	Signature	Value	i.
addObserver	(string,method)		
attributeWatchable	(string) : boolean		
Availability	real	100	i
AvailabilityOn	boolean	true	

The column signature allows you to deduce whether it is a method or an attribute and which data you need to pass or what type is returned. If the column only shows the data type, the entry then is an attribute.

Example: Show the attributes and methods of the store.

RecoveryTime	time	0.0000	i

Recovery time is an attribute; the data is not in parentheses. To set the recovery time, you have to type:

```
Store.recoveryTime:=120;
```

Set the value of an attribute with „:=".

PE	(integer,integer) : object

PE is a method. The method PE expects two arguments of data-type integer and returns an object. PE allows you to access a particular place of the store. You will call the method with parentheses:

```
Store.PE(1,1);
```

mirrorX
mirrorY

The row mirrorX does not contain a value in the column signature. MirrorX flips the icon on the x-axis. It is a method which has no arguments and returns no value. You will call this method without parentheses.

```
Store. mirrorX;
```

startPause	([time])

The parentheses in the column signature indicate that startPause is a method. The data-type is given within square brackets. This means that the argument is optional. This results in two possibilities of the call:

```
Store.startPause;   -- no time limit
Store.startPause(120);  -- pause for 120 seconds
```

Note:
Some attributes are read-only. You can assign no value to these attributes. Online help describes whether an attribute is read-only.

5.2.5   Templates

For a number of cases, Plant Simulation includes templates, which you can insert into your source code, or which you can use as a starting point for developing controls.

You reach Templates via the menu **TEMPLATE**:

Template	View	Tools	Help
if...then			Strg+Umschalt+I
if...then...else			Strg+Umschalt+E
if...then...elseif			Strg+Umschalt+S

In the method editor click into the row in which the snippet should be inserted. Then click, e.g., TEMPLATE – IF ... THEN ... Under SELECT TEMPLATE you find more templates. Watch out! Most templates in this selection completely replace your source code. You can use the Tab key for moving through the template (between the areas in angle brackets).

5.2.6 The Debugger

The debugger helps you to correct your methods. Using the F11 key, you can quickly change between editor and debugger (or RUN – DEBUG).

Example: Open the method from the example above, place the cursor in the text, and then press F11.

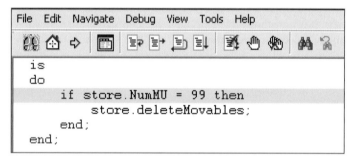

Using F11 you can, for example, move through your method stepwise and see what happens. If the method is completed, the method will open again in the editor. If an error occurs during the simulation, the debugger automatically opens and displays error messages.

Note:
If you have to make changes in the source code in the debugger, then you can save the changes by pressing the F7 key.

5.3 SimTalk

SimTalk does not differentiate between upper- and lowercasing in names and commands. At the end of a statement, you need to type a semicolon. Blocks are bounded by an "end" (no curly brackets as in Java or C++). If you forget the "end", Plant Simulation always looks for it at the end of the method.

5.3.1   Names

Plant Simulation identifies all objects and local variables using their name or their path. You can freely select new names with the exception of a few key words. The following rules apply:

- The name must start with a letter. Letters, numbers, or the underscore "_" may follow.
- Special characters are not allowed.
- There is no distinction between capital and lowercase letters.
- Names of key words and names of the built-in functions are not allowed.

For methods some names are reserved:

- **Reset**: Will be executed when clicking Reset in the Eventcontroller.
- **Init**: Will be executed when clicking Init in the Eventcontroller.
- **EndSim**: End of simulation (reaching the end time for simulation).

Generally, key words from SimTalk are prohibited names (all terms in SimTalk, which are highlighted in blue).

5.3.2   Anonymous Identifiers

SimTalk uses anonymous identifiers as wild cards. When running the method, these anonymous identifiers will be replaced by real object references.

root
The anonymous identifier "root" always addresses the top of the Frame hierarchy. Starting from this Frame, you can access underlying elements.

Example 36: root
First, open the console in Plant Simulation:
VIEW – TOOLBARS AND DOCKING WINDOWS – CONSOLE

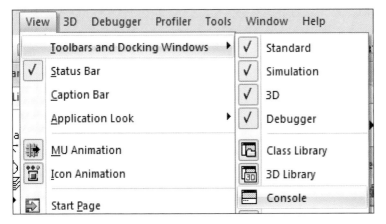

Create a new method. Use the method from the introduction example.
Complete the method as follows:

```
is
do
   -- writes the root name in the console
   print root.name;
end;
```

The console will show the name of the Frame in which the method was placed.

self
Self returns a reference to the current object (itself).

Example 37: Anonymous Identifier self

```
is
do
   -- The name of the current method will be
   -- written to the console
   print self.name;
end;
```

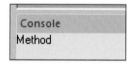

current
Current is a reference to the current Frame.

?
? denotes the object that has called the method (e.g., the object in which the method is entered as an exit control). The question mark allows a method to be used without modification by several objects.

@
@ refers to the MU which has triggered the method (so you can access, e.g., on all outgoing MUs).

basis
Basis denotes the class library. You can only use it in comparisons.

5.3.3   Paths

When objects are not located in the same frame or folder (name space), a path has to be put in front of the name. Only the path allows clearly identifying an object, so that it can be reached. A path is composed of names and periods (which serve as a separator). Paths are divided into two kinds of paths:

- Absolute paths
- Relative paths

5.3.3.1 Absolute Path

 The starting point of the absolute path is the root of the class library. From here on objects are addressed to the "bottom". An absolute path always starts with a period.

Example:

```
Modelle.Frame.workplace_1
```

5.3.3.2 Relative Path

A relative path starts within the frame, within which the method is located (without the first period).

Example:

```
workingplace_1
```

Workingplace_1 is located in the same Frame as the method.

```
controls.Method1
```

This address refers to an object with the name "Method1" in a subframe "controls".

5.3.3.3 Name Scope

All objects, which are located in the same frame or folder, form a name scope. Within a name scope identical names are not allowed. In other words, names in a name scope may occur only once, all objects must have different names. In different Frames identical names may occur. Their path distinguishes the objects. If you try to assign a name twice, Plant Simulation shows an error message:

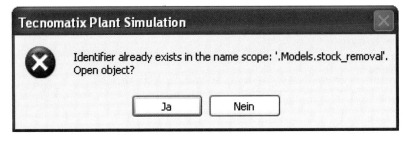

5.3.4   Comments

Comments explain your source code. Plant Simulation distinguishes between two
types of comments:

```
-- Comment until the end of line
/* Beginning of a comment, which
   extends over several lines
*/
```

Plant Simulation displays comments in green. We recommend to comment your
source code. In this comment, you should enter pertinent information about your
method. Plant Simulation provides a template for this purpose: **TEMPLATE –
SELECT TEMPLATE …**

You can find the header comment in **CODE SNIPPETS.**

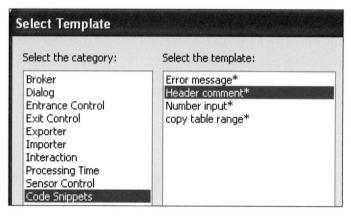

The header comment (started with "--" in front of the keyword is) is shown as a
Tooltip in the Frame, when you place the mouse on the method.

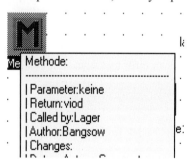

5.4   Variables and Data Types

5.4.1   Variables

A variable is a named location in memory (a place where the program stores information). Variables should have meaningful symbolic names. You first have to declare a variable (introduce its name) before you can use it. Plant Simulation distinguishes between local and global variables. A local variable can only be accessed within the respective method. This variable is unreachable for other methods. Global variables are visible for all methods (in every Frame). You can set values and get values from all methods. This way, you can organize data exchange between the components.

5.4.1.1   Local Variables

Local variables are declared between "is" and "do". A declaration consists of the name of the local variable, a colon, and a data type. The keyword "do" follows after the last declaration.

Example:

```
is
   Name  :  Type;
do
-- statements
end;
```

For instance:

```
is
   stock_in_store : integer;
do
… -- statements
end;
```

The declaration reserves an address in the memory and you define access on it by a name. For this reason, the operating system must know what you want to save. A true/false value requires less memory (1 bit) as a floating-point number with double precision (min. 32 bit). The information about the memory size takes place through the so-called data types. The data type determines the maximum value range of variables and regulates the permissible operations.

A value is assigned to a variable with :=.

Example 38: Declarating Variables

The circumference of a circle is to be calculated from a radius and Pi. Pi is defined in Plant Simulation and can be retrieved via Pi. The result is written on the console with the command print.

```
is
  radius :integer; --integer number
  circum :real; --floating point number
do
  radius:=200;
  circum:= radius*PI;
  print circum;
end;
```

SimTalk provides the following data types: acceleration, any, boolean, date, date-time, integer, length, list, money, object, queue, real, speed, stack, string, table, time, timeSequence, and weight.

Name	Range of values
acceleration	real, m/s²
any	the data type will be determined only after the assignment of the value (like VB: variant)
boolean	TRUE or FALSE
integer	-2.147.483.648 bis 2.147.483.647
real	floating-point numbers
string	character (each letter, numbers)
object	Reference to an object (except comment)
table	local variable with the behavior of a table
list	see above
stack	see above
queue	see above
money	...
Length	as real, the value is interpreted as meters
weight	see above, kg
speed	real → m/s
time	real → sec.; output: <hh>:<mm>:<ss.sss>
date, datetime	date from 1.1.1970 to 31.12.2038

Data types can be converted to a limited extent. Plant Simulation initializes all local variables automatically. The value depends on the data type:

Type	Initialization
boolean	FALSE
integer	0
real	0
string	"" (empty string)
object	Void
table	Void
list	Void
stack	Void
queue	Void
money	0
length	0.0
weight	0.0
speed	0.0
time	0:00:00
date	1.1.1970 0:00:00

If you want to define another start value in the simulation, you can, for example, use the init method.

Example 39: Global Variables

You need two methods in a Frame (Method1 and Method2):

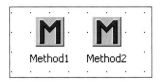

In Method2, define a variable of type integer with the name "number". Assign the value 11 to the variable.

Method 2:

```
is
    number:integer;
do
  number:=11;
end;
```

In Method1, you now try to read the variable "number" and to write the value of "number" to the console.

Method1:

```
is
do
```

```
print number;
end;
```

If you run Method1 (F5 or Run – Run), you get an error. It opens the debugger and the faulty call is highlighted. The error description is displayed in the status bar of the debugger:

Method1 cannot access a variable "number" of Method2.

If you need the data in several methods, you have to define a global variable (variable object from the folder information flow 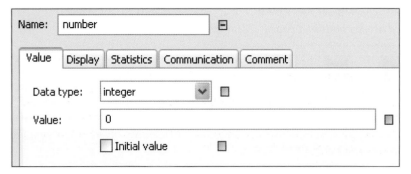 Variable) to exchange data. All methods can set and get the value of these variables. The global variable is defined as an object in the class library and is addressed just like the other objects by name and path. You have to determine the data type of the variable and can specify a start value.

Example 40: Global Variable 2

Insert a global variable into the Frame above. Rename the variable "number", data type integer (start value remains 0).

Change Method2 like this: Delete the variable declaration of "count", leave the rest unchanged:

```
is
do
   count:=11;
end;
```

Now start Method2 (the value of the global variable would have to change), and then Method1. The value of the global variable is displayed in the console.

Global variables are reset to the start value when you press the **RESET** button, if you select the option **INITIAL VALUE** and specify a start value. In the previous example, the value will be set to 0 when you enter the following setting:

5.5 Operators

By a combination of constants and variables, you can define complex expressions (e.g., calculations). Operators are used for concatenating expressions. Plant Simulation differentiates between:

- Mathematical operators
- Logical (relational) operators
- Assignment operators

Logical Operators are, for example, needed for comparisons.

5.5.1 Mathematical Operators

SimTalk recognizes the following mathematical operators:

−	Algebraic sign, subtraction
*	Multiplication
/	Division
//	Integer division
\\	Modulo (remainder of integer division)
+	Addition/concatenation of strings

The integer division, which is defined for the data-type Integer, always delivers a whole number. Any decimal places are suppressed. When calculating data for the data type real, the result is output up to seven valid digits (eighth digit rounded, working with decimal power).

5.5.2 Logical (Relational) Operators

Operator	Function	Result
AND	logical AND	TRUE, if all expressions are TRUE
OR	logical OR	TRUE, if at least one expression is TRUE
NOT	Not	Invert the boolean-value
<	Less than	
<=	Less than or equal	

>	Greater than	
>=	Greater than or equal	
=	equal	
/=	unequal	

A logical expression is always interpreted from left to right. The evaluation is completed once the value of an expression is established.

Example 41: Logical Operators

Simple method (+ start the console)

```
is
   local
      num1:integer;
      num 2:integer;
      num 3:integer;
      val1:boolean;
      val2:boolean;
      val3:boolean;
do
   num1:=10;
   num2:=23;
   num3:=num1*num2;
   val1:=num1<num2;
   val2:=num1<num2;
   val3:=val1 AND val2;
   print val3;
end;
```

Try out the operators. Let the console show different variables (Save + F5).

5.5.3   Assignments

The operator ": =" assigns a new value to a variable. First, the expression to the right of the operator is calculated. If the value and the variable have the same data type, then the value is assigned to the variable.

Example 42: Variable – Value Assignment

```
is
   num:integer;
do
   num:=1;
```

```
   num:=num+1;
-- first num+1, then assignment to num
   print num;    -- print the new value of num
end;
```

If the data types are different, the values have to be converted. Plant Simulation automatically converts real into integer and vice versa. For other types of data, you need to use type conversion functions.

Example 43: Type Conversion 1

```
is
   num:integer;
   text:string;
   res:integer;
do
   num:=10;
   text:="20";
   res:=num*text;
   print res;
end;
```

Run time error:

```
   text:="20";
   res:=num*text;
   print res;
```

Type mismatch.
stockRemoval

The most important type conversion functions are:

Syntax	Return value data type
bool_to_num(<boolean>)	real
num_to_bool(<integer>)	boolean
str_to_bool(<string>)	boolean
str_to_date(<string>)	time
str_to_datetime(<string>)	datetime
str_to_length(<string>)	length
str_to_num(<string>)	real
str_to_obj(<string>)	object

str_to_speed(<string>)	speed
str_to_time(<string>)	time
str_to_weight(<string>)	weight
To_str(<any>, …)	string

In the example above, a conversion from string to integer is required. You can realize a type conversion of text with the help of the function str_to_num (…). The method has the following syntax:

```
str_to_num (text)
```

Expand the example:

```
is
  num:integer;
  text:string;
  res:integer;
do
  nun:=10;
  text:="20";
  res:=num*str_to_num(text);
  print res;
end;
```

5.6   Branching

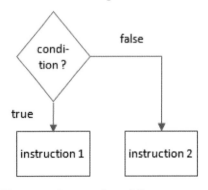

After testing a condition, the branch decides which of the following instructions should be executed. If the condition is met, the if-branch (TRUE) will be executed. If the condition is not met, the else branch (False) will be executed.

The general syntax is as follows:

```
if  condition  then
   instruction1;
else
   instruction2;
end;
```

Example 44: Branch 1

You only need one method for the example. We want to query if value1 is less than 10. If yes, then a message "If branch is executing" should be displayed in the console, otherwise "Else branch is executing"

```
is
  local
    value1:integer;
do
  value1:=12;
  if value1< 10 then
    print " If branch is executing";
  else
    print " Else branch is executing";
  end;
  print " Here we continue normally";
end;
```

Try different queries!

After passing through the branch, the execution of the code continues. If more than one condition is to be checked, the conditions can be nested. The nesting depth is not limited. In this case, a new condition begins after the if-branch with "elseif".

Example 45: Branch 2

Extension above:

```
is
  local
    value1:integer;
do
  value1:=7;
  if value1= 10 then
    print " value1 is 10.";
  elseif value1=9 then
    print " value1 is 9";
  elseif value1=8 then
    print " value1 is 8";
  elseif value1=7 then
    print " value1 is 7";
  -- and so on …
  end;
  print "Here we continue normally. ";
end;
```

```
Console
value1 is 7
Here we continue normally.
```

If you have to check many conditions, this construction gets complicated quickly. You can then use so-called case differentiations.

5.7   Case Differentiation

Case differentiation in SimTalk has the following syntax:

```
inspect <expression>
   WHEN <constant_1> THEN <instruction 1>
   WHEN <constant_2> THEN <instruction 2>
-- ...
end;
```

Example 46: Case Differentiation

```
is
   local
      num:integer;
do
   num:=2;
   inspect num
      when 1 then print "Num is 1.";
      when 2 then print "Num is 2.";
      when 3 then print "Num is 3.";
```

```
  -- and so on
  else
     print "Not 1, not 2, not 3 !";
  end;
end;
```

5.8 Loops

5.8.1 Conditional Loops

5.8.1.1 Header-Controlled Loops

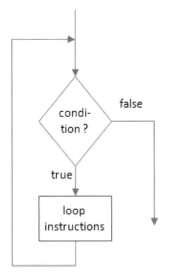

Before passing through the loop instructions, Plant Simulation checks whether a condition is met or not. The loop is repeated if the validation of the condition returns true. If the condition before the first loop is not met, the loop instructions will not be executed.

Make sure that the loop condition is false some of the time (e.g., increase in the value of a variable, until their value exceeds a certain limit).

Endless loops are terminated with the key combination CTR + ALT + SHIFT.

Syntax:

```
while <condition> loop
<instructions>
end;
```

Example 47: while-Loop

Loop 1, Loop 2 to Loop 10 should be written to the console.

```
is
   i:integer;
do
   i:=1;
   while i<10   loop
      print "Loop run number:" + to_str(i);
```

```
      i:=i+1;
    end;
  end;
```

5.8.1.2 Footer-Controlled Loops

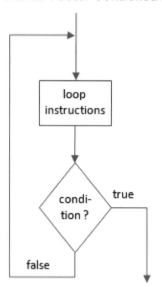

The condition is checked after the execution of the loop statement. If the condition for a termination of the loop is not met, the loop statement will be executed one more time. The loop statement is executed at least once.

Syntax:

```
repeat
  -- instructions
until <break condition is met>;
```

Example 48: repeat-Loop

```
is
  i:integer;
do
  i:=1;
  repeat
    print "Loop number:" + to_str(i);
    i:=i+1;
  until i>5;
end;
```

5.8.2 For-Loop

If you know exactly how often the loop is to be iterated, you can use the for-loop or the from-loop. It needs a running variable to control the number of runs of the loop. The variable is increased or decreased during each run starting from an initial value by a certain value. When a predetermined end value is reached, the loop will be terminated.

Syntax1:

```
from < Initialization > until <condition> loop
    <instructions>
end;
```

Syntax 2:

```
for < Initialization > to
<end value> loop
    -- loop instructions
next;
```

Example 49: from-Loop

Outputs will be shown in the console (loop 1, loop 2, and so on)

```
is
   local
   i:integer;
do
   from i:=1; until i=10 loop
     print "loop " + to_str(i);
     i:=i+1;
   end;
end;
```

```
Console
loop 6
loop 7
loop 8
```

Example 50: for-Loop

A loop should be executed 5 times:

```
is
   i:integer;
do
   for i:=1 to 5 loop
     print "loop " + to_str(i);
   next;
end;
```

You can count in the loop backwards with "downto" instead of "to".

Example 51: for-Loop with downto

Similar to the previous example:

```
is
   i:integer;
do
   for i:=5 downto 1 loop
     print "loop " + to_str(i);
   next;
end;
```

```
Console
loop 5
loop 4
```

5.9   Methods and Functions

A function is the definition of a sequence of statements, which are processed when the function is called. There are functions in different variants:

Arguments are passed to some functions, but not to others. (Arguments are values, which must be passed to the function, so the function can meet its purpose.) Some functions give back a value, others do not.

5.9.1   Passing Arguments

Arguments serve the purpose for passing data on during the function call (not just a stock removal function call, for example, but at the same time the number of the average stock removal per day is handed over by function call). The data type for the given value must match the data type of the argument declared in the function. The arguments have to be declared in the function. The declaration will be made at the beginning of the method before "is" in parenthesis in the following format:

```
(name : type)
```

For example:

```
(Stock_removal : integer)
```

Within the method, the arguments can be used like local variables, with the caller determining the initial values.

Example 52: Passing Arguments 1

The user is to enter the radius of a circle, and the size of the circumference is to be displayed in the console. To enter the argument, you need a text box. This is called by the function "prompt". Using the method "prompt", you can ask the user for input. If you pass a string to the method "prompt", then this string will be shown as a command prompt:

Names: Frame: Programming, Method: Test, the type of the data to be read is string, a type conversion is therefore necessary (str_to_num (identifier)).

Method Test:

```
is
   radius :string;
   circumference : real;
   val:real;
do
   -- prompt
   radius:=prompt("Radius");
   -- type conversion to real
   val:=str_to_num(radius);
   -- calculate circumference and display in
   --the console
   circumference:=val*2*PI;
   print    circumference;
end;
```

Open the console to see the result.

5.9.2   Passing Several Arguments at the Same Time

Within the definition, a semicolon separates several arguments. When you call the function, you have to pass the same number of arguments that you have defined in the function.

Example 53: Passing Arguments 2

For two given numbers, the larger number is to be returned after the call of the function. Name the function getMax (number1, number2).

```
(number1:integer;number2:integer)
-- passing arguments
:integer --type of the return value
is
do
   if number1 >= number2 then
      result:=number1;  -- return value
   else
      result:=number2;
   end;
end;
```

Call the function (from another method):

```
is
do
   print getMax(85,23);
end;
```

5.9.3   Result of a Function

Methods can return back results (usually a method that returns a value is called a function). For this purpose, you have to enter a colon and the type of the return value before "is". The result of the function has to be assigned within the function to the automatically declared local variable "result". Another possibility is to use "return". Return passes program control back to the caller. You can also pass a value on this occasion. After processing, the function will return the content of the variable "return" to the caller (the return value will replace the function call).

Example 54: Results of a Function

A function "circumference" is to be written. The radius will be passed to the function and the function returns the circumference.
Function circumference:

```
(radius:real) -- argument radius (2)
:real -- data type of the return value
is
do
result:=radius*2*PI; -- (3)
end;
```

Method Test (in the same Frame):

```
is
  res:real;
do
  res:= circumference (125); -- (1)
  print res;
end;
```

Explanation:

(1) A value is given when calling the function.
(2) The function declares the arguments.
(3) The function inserted the value passed at the designated position.

If you want to return more than one result from one function, result will not work. One solution is to define arguments as a reference. Usually the program makes copies of the data, and the function continues to work with the copies (except if you are passing objects; objects will always be passed as reference to the object). The original values of simple data types remain unchanged in the calling method. If arguments are defined as reference, you can change the values in the calling method by the called function.

The definition is accomplished with

```
(byref name:data_type)
```

5.9.4   Predefined SimTalk Functions

SimTalk has a range of ready functions.

5.9.4.1 Functions for Manipulating Strings

Function	Description
copy(<string>,<integer1>, <integer2>);	The function "copy" copies a number of characters (integer2) from a string starting from the position integer1. The first character has the position 0.
incl(<string1>, <string2>,<integer>);	The function "incl" inserts a string2 into the string1 before the position integer. The new string is returned.
omit (<string>,<integer1>, <integer2>);	"omit" copies the string and deletes the substring within it from starting position integer1 with number (integer2) characters. The new string will be returned.

pos(<string1>,<string2>)	"pos" shows the position within string2, in which the string1 occurs in for the first time. If the string1 is not contained in string2, 0 is returned, otherwise the position as integer.
strlen(<string>)	"strlen" returns the length of the string passed.

Example 51: Functions for Manipulating Strings

The file extension of a filename is to be identified and re-turned. The dot will be searched for first. Then, starting from a position after the dot all characters until the end of the string will be copied. The result will be displayed in the console.

```
is
   filename, extension:string;
   length, posPoint:integer;
do
   filename:="samplefile.spp";
   -- search point
   posPoint:=pos(".",filename);
   -- find out the length of the text
   length:=strlen(filename);
   -- copy the substring
   extension:=copy(filename,posPoint+1,length-
posPoint);
   -- display in the console
   print extension;
end;
```

5.9.4.2 Mathematical Functions

Function	Description
sqrt(x)	Square root. The argument x has to be greater than or equal to 0.
abs(x)	Absolute amount of x.
round(x)	Rounds x to the nearest whole number (on or off).
round(x,y)	Rounds x on y digits.
floor(x)	Nearest whole number less than or equal to x
ceil(x)	Nearest whole number greater than or equal to x

min(x,y)	Minimum
max(x,y)	Maximum
pow(x,y)	x^y
...	

5.9.5 Method Call

Methods are called by their names. During the simulation, methods have to be called often in connection with certain events.

5.9.5.1 Sensors

Most objects have sensors that are triggered when an MU enters the object and ex-its again. Length-oriented objects, such as the Line, have separate sensors for moving forward or in reverse.

Example 56: Methodcalls by Sensors

Parts manufactured by a machine are to be counted using a global variable. Create the following simple Frame:

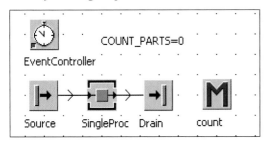

"COUNT_PARTS" is a global variable of type integer. The method "count" should look like this:

```
-------------------------------------------------------
--| increases the global variable "COUNT_PARTS" with
--| each call by one
-------------------------------------------------------
-
is
do
   COUNT_PARTS:= COUNT_PARTS +1;
end;
```

The method should now be called if a part is exiting the machine. Open the dialog of the SingleProc and select the tab CONTROLS.

Entrance:			...	□			
Exit:	count		...	▣	□ Front ▣	☑ Rear	▣

The main sensors are entrance and exit control. The setup control is triggered when a setup process starts and ends. You can activate the sensor with the front or the rear. The choice depends on the practical case. The rear exit control is triggered when the part has already left the object. For counting, this is the right choice. If you would trigger the control with the front, the MU will still be on the object. If the subsequent object is faulty or is still occupied, the front sensor is triggered twice (once when trying to transfer, the second time when transfer is done). If you select "front" in the exit control, then you need to trigger the transfer by SimTalk, even if a connector leads to the next object (e.g., @.move or @.move(successor)). If you press the F2 key in a field in which a method is registered, Plant Simulation will open the method editor and will load the relevant method.

5.9.5.2 Other Events for Calling Methods

For calling methods, you can use other events. These events can be found in the object dialogs under the command **TOOLS – SELECT CONTROLS** ...

Fail:		...	□
Pause:		...	□
Unplanned time:		...	□

The failure control, for example, is called if a failure of the object begins and ends (when the value of the property failure changes).

Example 57: Fail Control

You are to simulate the following part of a production. On a multilane, non-accumulating line parts are transferred and transported by 3 meters. Processing on each lane by a separate machine follows. If a machine fails, the entire line must be stopped (all lanes have to be stopped).Create the following Frame:

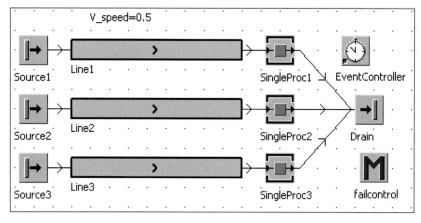

Settings: Source1 to Source3: interval 1 second, not blocking, Conveyors: 12 m, 0.5 m/s, SingleProc: each 1 second processing time, Failures: 90% availability, 10 minutes MTTR, Drain: 0 seconds processing time.

One way to stop all lanes by one machine failure could be: The method "failcontrol" checks whether a machine is failed. If at least one machine is failed, the speed of all three lines is set to 0. If no machine is failed, the speed is set to the value that is stored in the global variable "v_speed".

Method failcontrol:

```
is
do
   if SingleProc1.failed or
      SingleProc2.failed or
      SingleProc3.failed then
       Line1.speed:=0;
       Line2.speed:=0;
       Line3.speed:=0;
   else
       Line1.speed:=speed;
       Line2.speed:=speed;
       Line3.speed:=speed;
   end;
end;
```

Now the method "failcontrol" must be called whenever a failure occurs, or when the failure ends. Click in the dialog of the SingleProc objects **TOOLS – SELECT CONTROLS**: *Enter the method "failcontrol" as the failure control.*

Repeat this for all three machines. The result can very well be proven statistically. The individual machines can in theory work 90% of the time. If the failures of the other machines lead to no more parts being transported by the conveyor, then the machines remain 20% of their processing time without parts in addition to their own failures. That means the machines can on average operate only 70% of the time. Run the simulation for a while (for at least 2 days). Then click the tab **STATISTICS** *in SingleProc1, 2, or 3.*

5.9.5.3 Method Call After a Certain Timeout

You can call SimTalk methods after a certain timeout. This can be useful if you need to trigger calls after an interval referred to an event.

Example 58: Ref-Call

In this simulation, the machine should send a status message (ready) 10 seconds before completion of a part (e.g., to inform a loading device). The message should be shown first as output in the console. Create the following simple frame:

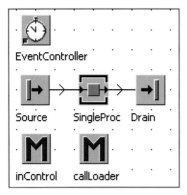

Settings: Source interval 2 minutes, SingleProc 2 minutes processing time, inControl as entrance control SingleProc. The method callLoader writes a short message into the console. Method callLoader:

```
is
do
    print time_to_str(eventController.simTime) +
    " called loader";
end;
```

The call of the method callLoader should now take place 10 seconds before completion of the part on the SingleProc. If you use the entrance control to start the timer, the method should be started after the processing time of SingleProc minus 10 seconds. For calling a method after a certain time period, you may use the method <ref>.MethCall(<time>,[<parameter>]) in Plant Simulation. You cannot address the method with a path or directly with name, because the method is then called directly. Instead, use the method ref (<path>). This returns a reference to the method by which you get access to the method. In the example above, the method InControl should look like this:

```
is
do
    ref(callLoader).methcall(
    SingleProc.procTime-10);
end;
```

6 Simtalk and Material Flow Objects

6.1 Attributes of the Material Flow Objects

All attributes that you can set in the dialogues of the material flow objects can also be set by SimTalk. This can be used, for example, to load the basic settings of the objects from a central location (e.g., a table). The Simtalk attributes are similar to the labels on the dialogs.

Important SimTalk attributes are:

- ProcTime (processing time)
- RecoveryTime (recovery time)
- CycleTime
- Capacity (if defined)
- MTTR
- MTTR.stream
- Availability

From version 9, you can create for each object a series of failures. You get access to the single failure on a failure listing (failures). The individual failures are addressed with her name. If your failure is, e.g., failure1, the failure is made accessible by failures.failure1. Each failure has the properties availability, MTTR, and MTTR.stream.

Example 59: Basic Settings

The data (processing time, recovery time, and one failure) of three objects will be read from a table when initializing the frame. Create the following frame:

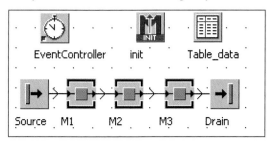

From version 9, you must create for M1, M2, and M3 first a failure. Name the failure in all three objects each "failure1". Format the table according to the following sample (take notice of the data types) and enter the data (see chapter Information flow objects – List Editor):

S. Bangsow: Manufacturing Simulation with Plant Simulation, Simtalk, pp. 117–138, 2010.
© Springer Berlin Heidelberg 2010

	object 1	time 2	time 3	real 4	time 5	integer 6	str 7
string	Machine	Processing_time	Recovery_time	Availability	MTTR	Stream	
1	M1	2:00.0000	10.0000	95.00	1:00:00.0000	9	
2	M2	1:30.0000	20.0000	80.00	30:00.0000	10	
3	M3	1:00.0000	10.0000	60.00	45:00.0000	11	

M1, M2, and M3 must have from version 9 a failure "failure1".

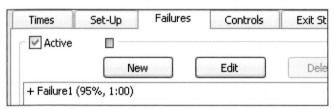

The init method should look as follows (see also chapter table):

```
is
     i:integer;
     machine:object;
do
     for i := 1 to table_data.yDim loop
       machine:=table_data["Machine",i];
       machine.procTime:=
       table_data["processing_time",i];
       machine.recoveryTime:=
       table_data["Recovery_time",i];
       /* version up to 8.2
       machine.availability:=
       table_data["Availability",i];
       machine.MTTR:=table_data["MTTR",i];
       machine.MTTR.stream:=table_data["Stream",i];
       */
       -- from version 9
       machine.failures.failure1.availability:=
       table_data["Availability",i];
       machine.failures.failure1.MTTR:=
       table_data["MTTR",i];
       machine.failures.failure1.MTTR.stream:=
       table_data["Stream",i];
     next;
   end;
```

6.2 State of Material Flow Objects

6.2.1 Operational, Failed, Pause

The method "operational" returns TRUE if the device is neither failed nor paused. Otherwise, the result is FALSE. With failed, you can determine if the object is currently in failure state or with pause, whether it is paused. Plant Simulation usually indicates the operating status of the objects with different colored LEDs. This state animation can be activated and deactivated in Plant Simulation with VIEW – ICON ANIMATION.

If you have to work with very small icons, then Plant Simulation shrinks the LEDs until they are irrecognizable. The best option is to change the whole icon, if the state of the object has changed. The following example gives a solution.

Example 60: Status Display of Material Flow Objects

To display the status of the material flow objects, we replace the icons of the objects. We realize the monitoring and controlling of the status display using an observer. You have to first create in the class SingleProc in the class library a series of icons for the various states. Make it in different colors. Use blue for pause (symbol number 2) and red for failed (symbol number 3). Other observable state are set up, unplanned and occupied (empty). Consider that the animation points of all icons are in the same position. Create the following frame:

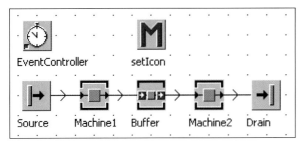

Settings:

Object	Attribute	Value
Source	Interval	2 minutes
Machine1, Machine2	Processing time	1 minute
	Availability	50%
	MTTR	30 minutes
Buffer	Capacity	100

After changing the properties, failed or pause of Machine1 should be called the method setIcon. This works best with an observer. Open the dialog of machine1 and select TOOLS – SELECT OBSERVERS.

Click Add and select as observed value FAILED as follows:

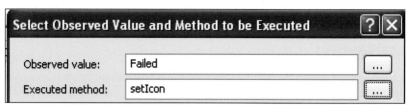

Analog you proceed for the attribute pause and Machine2. In the method CurrIconNo, now you must load the correct icon for the state of the machine.

- *Neither failed nor paused: Icon 1*
- *Paused: Icon 2*
- *Failed: Icon 3*

The calling machine you can request with „?". In this example, the method should look as follows:

```
(attribute: string; oldValue: any)
is
do
    if ?.pause then
          ?.CurrIconNo:=2;
    elseif ?.failed then
          ?.CurrIconNo:=3;
    else
          ?.CurrIconNo:=1;
    end;
end;
```

Example 61: Replacement Machine

Station_1 usually delivers parts to Station_2; If Station_2 is failed, Station_1 should deliver to replacement_machine.

Create the following Frame:

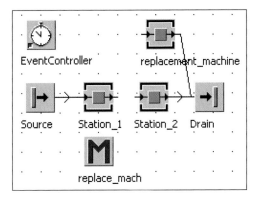

Method replace_mach → exit control front Station_1

```
is
do
   if  Station_2.operational=false then
      @.move(replacement_machine);
   else
      @.move(Station_2);
   end;
end;
```

Activate failures during the simulation for Station_2. The parts are redirected to the replacement machine.

6.2.2   Ready

The method returns TRUE if the object is occupied and an MU is ready to exit. *Syntax:*

```
<Path>.ready;
```

Unfortunately, there is no observable attribute that could be used to trigger an action, if the machine has finished processing and the part is ready for exit (e.g., to request a transporter). If you need, therefore, an observable attribute, you could use the following construction. Insert into the class of the object (e.g., SingleProc) a user-defined attribute (processingReady, datatype boolean, value false). Your user-defined attribute (data type boolean) is observable.

Name	Signature	Value	i.	Watchable
processingReady	boolean	false	ni	*

The attribute should be set at the entrance of a MU to false and then trigger the exit sensor to true. The methods could look as follows.

Entrance control:

```
is
do
    ?.processingReady:=false;
end;
```

Exit control:

```
do
    ?.processingReady:=true;
end;
```

You can monitor this attribute within a waituntil statement or by an observer.

6.2.3   Empty

The method "empty" returns TRUE if no MU (neither wholly nor partially) is located on the relevant object. You can also query certain places, e.g., of the ParallelProc (with x–y coordinates).

Syntax:

```
<path>.empty;
<path>[number1, number2].empty;
```

Example 62: Object State Empty

A buffer in the production is to be modeled. The buffer should be filled only if "sales problems" exist. When the buffer is empty, the part will immediately be transported from mach1 to mach2. When the buffer is not empty, the parts are redirected through the buffer to ensure the First in First Out principle.

Create the following Frame:

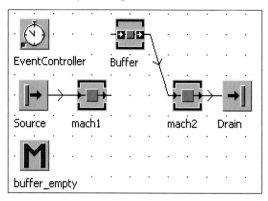

Settings: Capacity of the Buffer: 100, Source: interval 2 minutes, mach1: processing time: 2 minutes, 100% availability, mach2: processing time: 10 seconds, 50% availability, MTTR 10 minutes, mach1 exit control front: method Buffer_empty

Method Buffer_empty:

```
is
do
  if mach2.empty and mach2.operational and
    Buffer.empty then
    @.move(mach2);
  else
    @.move(Buffer);
  end;
end;
```

6.2.4   Occupied

The method "occupied" returns TRUE if the object contains at least one MU (either wholly or partially), otherwise it returns FALSE.

Syntax (analogous to empty):

```
<path>.occupied;
<path>[x,y].occupied;
```

6.2.5   Full

The method "full" returns TRUE if the capacity of the object is exhausted. For a place-oriented object, the method "full" returns TRUE if any of its places are occupied. If the object has several places, then each place can be queried individually (x–y coordinates).

Syntax:

```
<path>.full;
<path>[x, y].full;
```

Example 63: Method full

If the main warehouse is full, the part is to be stored in an alternative warehouse:
Method store (exit control object incoming).
   Create the following Frame:

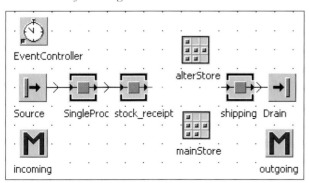

Method incoming; exit control object stock_receipt:

```
is
do
  if mainStore.full then
    @.move(alterStore);
  else
    @.move(mainStore);
  end;
end;
```

Number of MUs

```
<path>.numMU
```

The method numMU returns the number of MUs booked on the object.

6.2.6   Capacity

```
<path>.capacity
```

The method "capacity" returns the number of places of an object. In addition, you can query all states (paused, failed, blocked ...) which an object can take.

Example 64: Machine with Parallel Processing Stations

We want to simulate parallel processing. Five parts will be mounted on a machine and then processed together within 20 minutes. After processing, the five parts exit the machine (almost) simultaneously. The machine receives one part every 4 minutes. The following approach would be possible: The parts will be collected in a store until the required number of parts is reached. If the number of parts is reached, the entrance of the store will be locked. If the machine is empty, then all the parts are moved onto the machine by a loop. After machining of all parts, the last part (exit control rear) removes the blockage of the entrance of the store. Then, the cycle starts again.

   Create the following Frame:

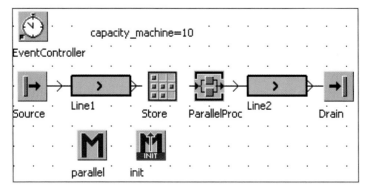

Method parallel:

```
is
   i:integer;
do
   if ?=Store then
   -- eventhandling of the store
     if store.numMU = capacity_machine then
       waituntil ParallelProc.operational and
               ParallelProc.empty prio 1;
       -- move all parts
       for i:=1 to capacity_machine loop
         store.cont.move(ParallelProc);
       next;
       -- lock the entrance of the store
       Store.entranceLocked:=true;
     end;
   elseif ?=ParallelProc then
     if ParallelProc.empty then
       Store.entranceLocked:=false;
     end;
   end;
end;
```

Explanation:

The control is to be used for two objects. With „?" you can query which object initiated the call. You can program the control with a respective conditional branch:

```
if  ? =  Store then
-- method is called from the store
elseif ? = ParallelProc then
-- method is called from the ParallelProc
end;
```

Method init:

Capacity_machine is defined as global variable, so that this object can be easily adapted (especially the capacity). This global variable will be read by the init method, and the capacity of the store and ParallelProc will be adjusted accordingly. You have to also clear the entrance of the store in case the simulation stopped during processing of the ParallelProc.

```
is
do
   deleteMovables;
   Store. entranceLocked:=false;
   -- check and possibly reset the value of
   -- the global variable
```

```
if capacity_machine <1 then
   capacity_machine := 1;
end;
--set the capacity of the ParallelProc
ParallelProc.ydim:=1;
ParallelProc.xdim:=capacity_machine;
--capacity of the Store
Store.ydim:=1;
Store.xdim:= capacity_machine;
end;
```

6.3   Suspending Methods

In the present case, transferring the parts to the next machine must wait until the machine is empty and the machine is neither failed nor paused (the transfer would run, not succeed).

To solve this problem, there are at least two ways:

- The execution of the method is suspended until the condition is met, then execution will continue.
- You can monitor the conditions for a transfer of parts, if all conditions are right, transfer will be triggered (Observer).

In the example above, we realized the first variant. The execution of methods can be interrupted for many reasons. One method calls another, for example, or it transfers an MU to an object with an entrance control. The execution of the method is interrupted as long as the new method is active, and will resume immediately after the method has finished. Often you have to wait for certain events in the simulation, for example, that a failure ends, processing is complete, a vehicle is available, etc. Until the occurrence of the condition, the method has to be interrupted.

Syntax:

```
waituntil <boolean expression> prio <gZ>;
```

The statement consists of the keyword "waituntil", a condition followed by the keyword "prio", and an expression to calculate the priority.

Condition: The expression specifies the conditions under which the execution of the method will continue. If the condition is met, the interpreter continues the execution with the following statement. If the condition is not met, then the interpreter suspends the method and monitors the various elements of the expression. If a part of the condition changes later, then the interpreter re-awakens the method so that the condition can be evaluated again. The structure of the expression is limited, because the interpreter must decompose the expression into observable components. Permissible are the basic arithmetic operations (+, −, *, /), comparisons

($<$, $<=$, $=,>$ $=,>$, $/ =$), and parentheses. Not permitted are method calls, table requests, and standard methods with arguments.

Priority: Several methods may have been suspended due to the same or to a similar condition and will therefore also be woken up together. The interpreter selects the method with the highest priority, and wakes this method up first.

Observable values: The condition on which the continuation of a method depends must be observable by Plant Simulation.

The following values are observable:

General: values of global variables and free attributes, which have "scalar" data types (boolean, integer, real, string, object, time, money, length, weight, speed, date, datetime).

6.4 Observer

Plant Simulation can watch observable values by themselves and invoke a method when changing a value. The name of the property (as a string) and value (before the change) will be passed. You can access the object itself with „?".

Example 65: Observer

For comparison reasons, we want to simulate the sample "Machine with parallel processing" with observers. Preliminaries: The conditions for transfer and locking of the entrance of the store are: The store is full, the ParallelProc is empty, and the ParallelProc is operational (not failed and not paused). If all conditions are met, all the parts are moved from the store to the ParallelProc; the entrance of the store is locked to prevent that parts are moved to the "working" Parallel-Proc. The method "parallel" should therefore look like this in the revised version:

```
(attrib:string;value:any)
is
   i:integer;
do
   if store.numMU = capacity_machine and
        ParallelProc.operational and
        ParallelProc.empty then
     -- move all parts
     for i:=1 to capacity_machine loop
       Store.cont.move(ParallelProc);
     next;
   --lock the entrance of the store
   Store.entranceLocked:=true;
   end;
end;
```

Unlocking the entrance of the store must be carried out in a separate method (e.g., method Store_unlock):

```
is
do
    if parallelProc.empty then
            Store.entranceLocked:=false;
    end;

end;
```

Assign the method "Store_unlock" as the output control of the ParallelProc. Remove the method "parallel" as the input control of the store. The objects whose properties you want to observe (once by the store, and twice by the ParallelProc) call the method, "parallel". Select **TOOLS – SELECT OBSERVERS** *in the dialog of the object Store.*

Click Add. You can enter the observed value or select it from a list. You have to watch the property numMU. The name of the method is "parallel".

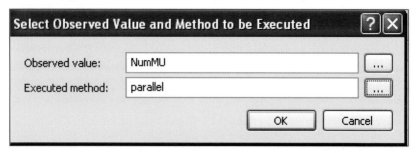

In the object ParallelProc, these are the attributes operational and empty.

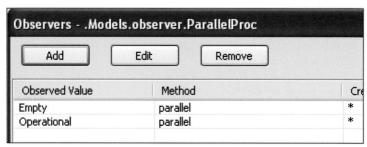

Start the simulation and test the behavior.

6.5   Content of the Objects

The following methods relate to the content of an object.

deleteMovables

```
<path>. deleteMovables
```

The method deleteMovables destroys all MUs which are booked on the object.

Example 66: deleteMovables

If you want to destroy all MUs in the Frame, then the following method is sufficient:

```
is
do
   deleteMovables;
end;
```

Cont (Buffer: MUPart)
Cont returns the MU, which is booked on the object. If there is no MU booked on the object, then the return value is VOID.

```
Syntax: <path>.cont;
```

Example 67: Method cont

The occupation of a machine is to be displayed in the console.
Create the following Frame:

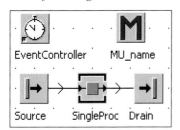

Assign the method MU_Name as the entrance control of the SingleProc.

Method MU_name:

```
is
do
   print SingleProc.cont;
end;
```

MU.name
MU.name returns the name of the MU.

Example 68: Surface Treatment

A number of different parts run through a chemical process (30 minutes surface treatment). After that, the parts are further processed on different machines. The simulation flow must branch out after the surface treatment. The following simple example will illustrate this.

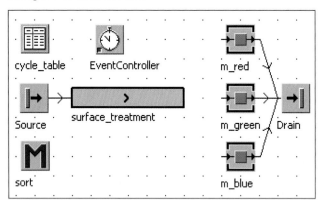

Settings: Source: interval 1 minute, sequence cyclical (cycle_table); a cycle consists of three parts red, two green, and two blue. Create three MUs in the path models: red, green, and blue for the simulation. Color the icons according to the designations.

cycle_table (after formatting):

	object 1	integer 2	string 3
string	MU	Number	Name
1	.MUs.red	3	
2	.MUs.blue	2	
3	.MUs.green	2	

surface_treatment: length 9 meters, processing time 30 minutes, M_red, M_green, and M_blue processing time each 1 minute, 100% availability.

Method sort (exit control (front) of surface_treatment):

```
is
do
   inspect @.name
   when "red" then
      @.move(m_red);
   when "blue" then
      @.move(m_blue);
   when "green" then
      @.move(m_green);
```

```
      end;
   end;
```

MU
Syntax:

```
<path>.MU(<integer>)
```

The method MU accesses all MUs whose booking points are located on the object. If no argument is given, then the first MU is returned. If the argument is greater than the number of booked MUs, then VOID is returned.

Note:
Use the method isVoid (value) to check whether a certain value is VOID (in the case of MU: if an MU is located on the object).

Example 69: Method MU

A list of all MUs in the store is to be displayed. Create the following Frame:

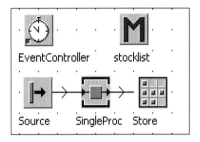

Method stocklist:

```
   is
      local
         i:integer;
   do
      from i:=1 until i > Store.numMU loop
         print Store.MU(i);
         i:=i+1;
      end;
   end;
```

Run the simulation for a while. Then call the method (right click – run).

numMU
Syntax:

```
<path>.numMU;
```

This method returns the number of MUs, which are booked on the object. The data type of the return value is integer.

6.6   Sensors

Lines and tracks can be very long. Therefore, it may be useful to trigger methods, if the MU is located some way before from the end, or to set few breakpoints at which methods should be executed. For this purpose, you can define user-defined sensors. The sensors act like switches. When a MU crosses (forward or backward) the sensor, the switch is activated and a method is called. In the method (control), you have to determine what happens at this position. A small example illustrates this:

Example 70: Sensors, Color Sorting

Within a production facility, parts with different colors arrive. With the help of cameras and sorting facilities, the parts will be distributed to different color-homogenous lines. We do not want to simulate the sorting facility.

Frame: main_line: 35 meters, L_red, L_blue and so on: 5 meters. All lines are accumulating, speed 1 m/s, no acceleration. All L_-lines are connected to drains. The drains have a processing time of 0 seconds. The source is connected to the main_line.

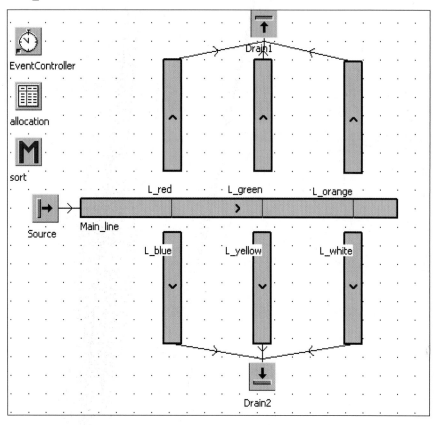

Duplicate six entities in the class library. Rename them to red, blue, green, yellow, orange, and white. Type in an icon size of 11 x 11 pixels and color the icons with the respective colors. The source should randomly produce these parts with a percentage of 16.7% each. Select the following settings in the source: interval 0.5 seconds, MU-Selection: random, table: allocation. Enter the following data into the table distribution:

	object 1	real 2	string 3
string	MU	Frequencies	Name
1	.MUs.red	0.17	
2	.MUs.green	0.17	
3	.MUs.blue	0.17	
4	.MUs.yellow	0.17	
5	.MUs.white	0.17	
6	.MUs.orange	0.17	

Note:
Drag the MUs from the class library to the table to enter the absolute paths of your MUs. Depending on the position of the MUs in the class library you may have other addresses in the table.

*Insert three sensors on the main_Line (10 m, 20 m, and 30 m each rear). Assign the method sort to all sensors. Proceed like this: Click the button **SENSORS** on tab control in the dialog sensor of the main_Line. Then, click the button **NEW**.*

Enter a position (e.g., 10 m). Decide whether the front or the rear (checkmark) of the MU should trigger the sensor. Select the control "sort".

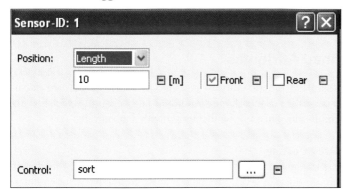

Complete the sensor list like this:

ID	Position	Front	Rear	Path
1	10m	x		sort
2	20m	x		sort
3	30m	x		sort

The method "sort" will check the name of the MU. If the MU "red" arrives at sensor 1, then the MU is to be moved to the line L_red, etc.

```
(sensorID : integer)
is
do
  if sensorID=1 then -- first sensor
  -- red to L_red, blue zu L_blue
    if @.name="red" then
      @.move(L_red);
    elseif @.name="blue" then
      @.move(L_blue);
    end;
  elseif sensorID=2 then
    if @.name="green" then
      @.move(L_green);
    elseif @.name="yellow" then
      @.move(L_yellow);
    end;
  elseif sensorID=3 then
    if @.name="orange" then
      @.move(L_orange);
    elseif @.name="white" then
      @.move(L_white);
    end;
  end;
end;
```

The parts should now be color-sorted on the L_-lines.

6.7   User-Defined Attributes

You can always extend the functionality of the objects with user-defined attributes. This way, you can take aspects in the simulation into account, which are not included in Plant Simulation. This will be illustrated with an example.

Production-Associated Costing
For many aspects it is interesting how much production cost a product already has caused up to a certain production level. Production costs are the basis for calculat-

ing fixed capital within a production process (e.g., current assets, inventory). The target of many improvement measures is to reduce fixed capital. For determining manufacturing costs, different ways of calculation exist. A possible calculation is:

Manufacturing materials
+ Manufacturing wages
+ Special direct costs of production
= Minimum production costs
+ General expense for materials
+ General expense for production
+ General expense for administration
+ Interest on debt capital
= Highest production costs

The general expense is calculated like this: direct expenses * cost rate. The following simple approach should be chosen for the simulation:
    The basis for determining the production costs are the costs of materials and direct manufacturing costs. Direct production costs are calculated using the time multiplied with the hourly wage or the hourly machine rate. The individual parts now must collect this information during processing (simulation). One way to realize this is to use user-defined attributes.

Example 71: Production Costs and Working Assets

Create the following Frame:

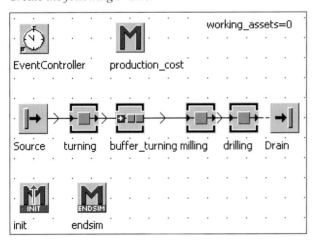

Material costs of unfinished part: €24.95, average manufacturing wage €/h36. First, the part is turned, then milled, and then drilled. Processing times: turning 1 minute, milling, drilling 1 minute. The relevant entity (Part) is to have the property "production_cost" (data type real).

Procedure: Duplicate an entity. Rename it to "part". Open the entity by double-clicking it in the class library. Select USER-DEFINED ATTRIBUTES – NEW.

| Attributes | Product Statistics | User-defined Attributes | | | |

| | | New | | Edit | | Delete | |

Name		Value	Type		C	I.	3D

Enter the following values in the dialog:

User-defined Attributes

Name: production_cost

| Value | Statistics | Communication |

Data type: real

Value: 24.95

The value of the attribute is initialized with the value of the raw material (at the beginning of processing). The machines need to have an attribute "wage", into which the hourly wage costs are entered.

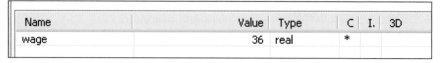

Name		Value	Type		C	I.	3D
wage		36	real		*		

The calculation of the production costs is relatively simple (e.g., exit control rear of the machines, method production_cost):

```
is
do
    @.production_cost:=
    @.production_cost+((?.procTime/3600)*(?.wage));
    -- procTime in seconds!
end;
```

Note:
The anonymous identifier ? returns a reference to the object whose control was triggered.

Even in assembly operations, the manufacturing costs of the parts can be combined and transferred to the new part (if you destroy the individual parts).

Working Assets
To determine the working assets, you have to identify the existing entities and their cost (ideally within an EndSim method).

Continue Example:

The value (costs) of the parts in the Frame are to be determined after the end of the simulation. For this purpose, you query the individual objects whether they are occupied or not. If they are occupied, then the costs of the parts are added to a global variable (working_assets).

Method endSim:

```
is
  i:integer;
do
  if turning.occupied then
    working_assets:=
    working_assets+turning.cont.production_cost;
  end;
  if milling.occupied then
    working_assets:=
    working_assets+milling.cont.production_cost;
  end;
  if drilling.occupied then
    working_assets:=
    working_assets+drilling.cont.production_cost;
  end;
  --query each place individually
  if Buffer_turning.occupied then
    from i:=1; until i=Buffer_turning.capacity loop
      if Buffer_turning.pe(i).cont /= void then
        working_assets:=working_assets+
        Buffer_turning.pe(i).cont.production_cost;
      end;
     i:=i+1;
    end;
  end;
end;
```

For a large number of objects this approach is very involved. With the Frame object, you can access all objects in the Frame. With Frame.numNodes you can determine the number of all objects in the Frame. The individual objects can be accessed with Frame.node(index). With the method class, you can check the type of an object. A universal method for calculating the working assets could look like this:

```
is
  i:integer;
  k:integer;
do
  working_assets:=0;
  for i:=1 to current.numNodes loop
    if current.node(i).class.name="SingleProc" or
       current.node(i).class.name="PlaceBuffer"
-- and so on
      then
        for k:=1 to current.node(i).numMu loop
            working_assets:=working_assets+
            current.node(i).mu(k).production_cost;
        next;
      end;
    next;
end;
```

7 Mobile Units

MUs represent the materials that flow from object to object within the Frame. After creating new MUs, they move through the model and remain at the end until they are destroyed.

7.1 Standard Methods of Mobile Units

7.1.1 Create

Syntax:

```
<MU_path>.create(<object>[,length]);
```

The method creates an instance of MU on <object>; MU-path is the path to the MU (e.g., class library). Optionally, you can specify a length on a length-oriented object (e.g., track, line). If you do not specify a length, then the MU is generated at the end of the object (ready to exit).

Example 72: Create MUs

At 200 m, 500 m, and 700 m, a transporter is to be created on a track which is 1000 m long.

Create the following Frame:

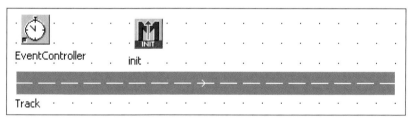

Note:
If you want to create track, which is a 1000 m long by dragging, then you first have to change the scaling in the Frame window. The default setting is a grid of 20 x 20 pixels and a scale of 1 m per grid point. Therefore, you have to change the scale so that it shows 50 m per grid point (Frame window, TOOLS – SCALING FACTOR …). Another possibility is the following: On the tab CURVE (track) turn off the option TRANSFER LENGTH.
Then you scale a length of 1000 m on the tab ATTRIBUTES. Then the track is no longer shown to scale.

S. Bangsow: Manufacturing Simulation with Plant Simulation, Simtalk, pp. 139 – 181, 2010.
© Springer Berlin Heidelberg 2010

The init method is to first delete all MUs and then to create the three transporters:

```
is
do
   deleteMovables;
   .MUs.Transporter.create(track,200);
   .MUs.Transporter.create(track,500);
   .MUs.Transporter.create(track,700);
end;
```

After creating the MU on a place-oriented object, it can exit immediately. On a buffer, MUs are created on the first free place, if no place has been specified. On a length-oriented, object the MU is created as close as possible to the exit if no position was specified. Creation will fail if the capacity of the object is exhausted and the length-oriented object is shorter than the MU to be created (Return value: VOID).

7.1.2 MU-Related Attributes and Methods

Method	Description
`<MU-path>.delete;`	This method destroys the specified MUs. It does delete MUs in the class library.

`<MU-path>.move;`
`< MU-path>.`
`move(<target>);`
`<MU-path>.move(<index>);`
	Move the front of the transferred MU. If no argument is specified, then it will be transferred to each successor alternately. If should be moved to a particular successor, then you can use an index (index). The return value (boolean) is TRUE when moving was successful and FALSE if moving failed (successor is occupied, paused, failed). MUs cannot be moved to a source.

`<MU-path>.transfer;`
`<MU-path>.transfer`
`(<target>);`
`<MU-path>.transfer`
`(<index>);`
	Transfer moves a MU from one object to another. It moves the entire length to the next object (not just the forefront like the method Move).

`<MU-path>.numMU;`	NumMu returns the number of MU, which are booked on the MU (integer).

Within an entrance or exit control, you can access the triggering MU with the anonymous identifier "@".

7.2   Length, Width, and Booking Point

All MUs have the properties length, width, and related booking points. The booking point determines the position of the MU, from this it is booked on the next object and can be accessed from there. The position of the booking point must be always less than or equal to the length or width. Otherwise, you get an error message. In some cases it is necessary to dynamically adjust the length during the simulation. This may be when the processing is changing the length or if you repeatedly change the direction of transport during the simulation.

Example 73: Change MU Length

We want to simulate a small part of a production. There are steel beams processed. The beams are initially transported lengthwise, then crosswise, then processed, and then transported again lengthwise. The cross conveyor serves as a buffer. Create the following frame:

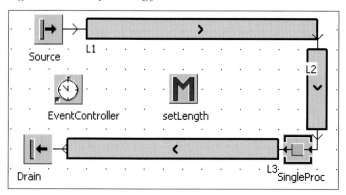

Create an entity ("beam") in the class library. The beam is 4 meters long and 20 cm broad. Set the booking points for each length and width to 0.1 meters. Redesign the icon of the beam: width 80 pixels, height 5 pixels. Delete the old icon and fill the icon with a new color. Set the reference point to the position 40.3. Change the icon "waiting" analogous. Make the following settings:

The source creates each 3 minutes one beam. The lines L1 and L3 are 12 meters long and transport the beams with a speed of 0.05 meters per second. Uncheck the option Rotate MUs in the tab Curve in L2 so that the line L2 transports the beams "cross". The SingleProc has 2 minutes processing time, 75% availability and 30 minutes MTTR. Start the simulation. If a failure occurs in SingleProc, the beams at L2 do not jam as intended. In the original setting, L2 promotes the cross transport, but the capacity is calculated on the length of conveyor line and length of the beam. If the beam is 4 meters long and L2 also exactly one beam fits on the line. To correct this, you have to temporarily reduce the length to the width of the beam (from 4 meters to 20 centimeters). After leaving the cross conveyor, you have to reset the length again to 4 meters. Therefore, insert two user-defined attributes into the class beam (class library):

Attributes	Product Statistics	User-defined Attributes		

New		Edit		Delete

Name		Value	Type	C
mu_length		4	real	*
width		0.2	real	*

The method setLength must be called twice, once at the entry of MUs in the line L2 (reduce length) and once when MUs enter the SingleProc (set length back to the original value). The method could look as follows:

```
is
do
    if ?=L2 then
        @.muLength:=@.width;
    elseif ?=SingleProc  then
        @.muLength:=@.mu_length;
    end;
end;
```

The beams jam now at L2 if a failure occurs at the SingleProc.

7.3   The Entity

The entity does not have a basic behavior of its own. It is passed along from object to object. The main attributes are length and width. Booking points determine at which position the entity will be booked while being transported to the succeeding object (mainly track and line).

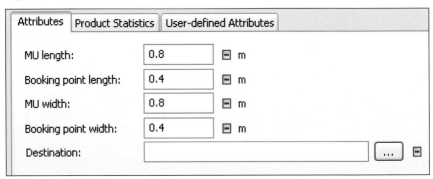

The destination can be used to store the destination station during transport operations (especially during transportation by a worker).

7.4 The Container

The container can load and transport other MUs. It has no active basic behavior of its own. The storage area is organized as a two-dimensional matrix. Each space can hold one MU. The container is transported from object to object along the connectors or by methods. With containers you can, for example, model boxes and palettes.

7.4.1 Attributes of the Container

The attributes of the container are mostly identical to those of the entity. In addition, the container has the following attributes:

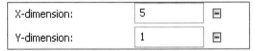

The capacity of the container is calculated by multiplying the X-dimension with the Y-dimension. The access to the MUs, which are transported by the container, is analogous to the material flow objects.

Method	Description
`<MU-path>.cont`	The method "cont" returns a MU, which is booked on the container (with the longest length of stay).
`<MU-path>.pe(x,y)`	With "pe" you get access to a place in the container.

7.4.2 Loading Containers

The approach is somewhat complicated. A container cannot exist just by itself. That is why you need another object, which can transport the container, e.g., to load a box. In addition, transporters can easily transport containers. For loading containers, you can use the assembly station (see chapter "AssemblyStation") or the transfer station. You can also load the container with SimTalk methods.

Example 74: Loading Containers

We want to develop a method that creates a palette that is loaded with 50 parts and to pass these parts onto the simulation. Create the following Frame:

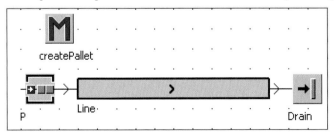

The method createPallet must first create the palette on P and then create parts on the palette until the palette is full. Create duplicates of container (Pallet, X-dimension: 25, Y-Dimension: 2) and entity (part) in the folder MUs.

Method createPallet:

```
is
do
   --create palette
   .MUs.pallet.create(p);
   -- create parts
   while not p.cont.full loop
       .MUs.part.create(p.cont);
   end;
   -- pass the palette
   p.cont.move;
end;
```

Digression: Working with Arguments 1
You will also need the facts described above in the exercises below. Therefore, it would be useful if the method could be applied to all similar cases. For this purpose, you need to remove all direct references to this particular simulation from the method (mark bold, italic).

```
is
do
   -- create palette
   .MUs.pallet.create(p);
   -- create parts
   while not p.cont.full loop
       .MUs.part.create(p.cont);
   end;
   -- pass the palette
   p.cont.move;
end;
```

The method only contains three specific references:

- Location of the palette (.MUs.pallet)
- Location of the part (.MUs.part)
- Target of creation (p)

These three items are declared as arguments.

```
(pallet,part,place:object)
```

Next, replace the specific details with the arguments (using find and replace in larger methods).

```
(pallet,part,place:object)
is
do
  -- create palette
  pallet.create(place);
   -- create parts
  while not place.cont.full loop
     part.create(place.cont);
  end;
  -- pass the palette
  place.cont.move;
end;
```

The method call now just has to include the arguments. Create an init method. The call of the method createPallet in the init method could look like this:

```
is
do
    createPallet(.MUs.pallet,.MUs.part,p);
end;
```

Start the init method. To check whether the method has really produced 50 parts, check the statistics of the palette (double-click the filled palette) on the tab STATISTICS.

Contents:	50
Minimum contents:	0
Maximum contents:	50
Entries:	50

7.4.3   Unloading Containers

You can easily simulate unloading processes using the dismantle station. For some simulation tasks, this approach might prove to be too cumbersome, and the number of objects would be growing enormously. You can easily program unloading the palettes with SimTalk. Accessing the palette and the parts on the palette takes place by using the "underlying" object.

Example 75: Batch Production

You are to simulate batch production. Parts are delivered in containers and placed close to the machines. The machine operators remove the parts from the container and place the finished parts onto another container (finished parts). Once the batch is processed, the container with the finished parts will be transported to the next workplace and the empty palette will be transferred. Create a Frame with a single machine:

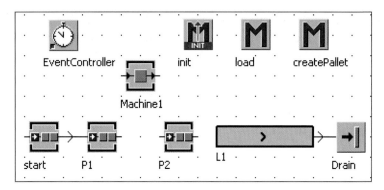

Settings: Start, P1, P2 each one place, processing time 0 seconds, L1 4 meters length, speed 1 m/s, Machine1 processing time one minute, no failures.

Digression: Reusing Source Code from Other Models
We want to use the same method (source code) as in the example "loading of containers" from the previous chapter. You can do this in two ways:

1. Save and import the method as an object file.
2. Export and import the methods as text.

Saving and importing the method as an object file:

You can store classes and Frames as objects at any time. The functionality is located in the class library. Therefore, you first have to create a "class" out of the method in the Frame. Hold down the Ctrl key and drag the method from the Frame to the class library (if you do not hold down the Ctrl key, the method will be moved!). If needed, rename the method in the class library. From here, you can save the method as an object. Select the context menu (right mouse button) and select SAVE OBJECT AS.

Now you can load the method into another file as an object file. Click a folder icon with the right mouse button. Select SAVE/LOAD-LOAD OBJECT INTO FOLDER... from the context menu:

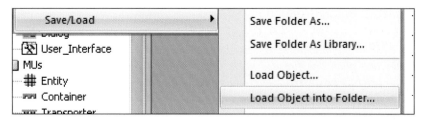

You can now select the file and insert the object into the class library. Caution: When you import the objects, the compatibility of versions/licenses will be considered. You cannot use objects from newer versions of Plant Simulation in older versions (or in eM-Plant).

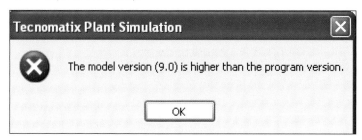

Exporting and importing a method as text:

You can export and import the source code of the method as a text file. Open the method createPallet. Select **FILE – SAVE AS** in the method editor:

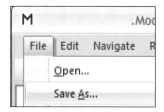

Select a location and a name for the file. Insert a method in the new Frame (without content, the contents will be completely replaced when you import a text file). Select **FILE – OPEN** in the method editor: Select the text file with the source code, and confirm. The text of the method is completely overwritten with the contents of the file.

Continuation Example Batch Production

If not already available, create the palette container (capacity 50 parts), entity: part. The init method should produce a full palette (50 parts) at the station P1 and an empty palette at the station P2.

```
is
do
   deleteMovables;
   createPallet(.MUs.pallet,.MUs.part,start);
   .MUs.pallet.create(p2);
end;
```

The simulation model (without a dismantle station) could look as follows. The palette itself transfers the first part of the palette onto the machine (exit control of P1). If a part on the machine is ready, it triggers the exit sensor of the machine. A

method transports the part to the palette on the station P2, and a new part from
the palette on P1 to the machine. If the finished parts palette is full (p2), they will
be transported to F1, the empty palette is moved from P1 to P2, and a new palette
will be generated at the beginning.

Method load (exit control front P1, exit control front machine):

```
is
do
  if ? = P1 then
  -- load the first part onto machine1
    @.cont.move(machine1);
  elseif ?=machine1 then
  -- load the part onto the palette
    @.move(p2.cont);
    if p2.cont.full then
      p2.cont.move(L1);
      p1.cont.move(p2);
      -- create a new palette
      createPallet(.MUs.pallet,.MUs.part,start);
    else
    -- already parts on p1
      p1.cont.cont.move(machine1);
    end;
  end;
end;
```

You can access the part on the palette with

```
p1.cont.cont.
```

In our case p1.cont is a palette and p1.cont.cont is a part.

Digression: Working with Arguments 2 – User-Defined Attributes
Imagine that the simulation of the production contains 50 machines (each with two
buffer places). You need to extend the method "load" for each machine. There is
an easier option though, namely to separate event handling of the buffer and the
machine.

Method: loadFirstPart (exit control front P1):

```
is
do
  -- move the first part to the machine
  @.cont.move(machine1);
end;
```

Method load (without branching after objects):

```
is
do
   -- load the part onto the palette
   @.move(p2.cont);
   if p2.cont.full then
     p2.cont.move(L1);
     p1.cont.move(p2);
     -- create a new palette
     createPallet(.MUs.pallet,.MUs.part,start);
   else
     -- already parts on p1
     p1.cont.cont.move(machine1);
   end;
end;
```

First step: Which object in the method has a reference to a specific case?

- P1, P2
- L1
- Machine1
- createPallet (only machine1)

The anonymous identifier "?"can replace Machine1. P1 and P2 are buffers which belong to the machine. If you are using many similar machines, it is worth your while to define the respective attributes in the class SingleProc (in the class library). Name the attributes:

- bufferGreenParts (object)
- bufferReadyParts (object)
- successor (object)

Open Machine1 in the Frame. In the dialog of the object select:

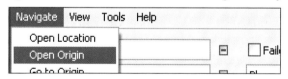

It will open the class of the SingleProc in the class library: Click the tab **USER-DEFINED ATTRIBUTES**, and define the two buffers and the successor:

Importer		Failure Importer		User-defined Attributes		
		New		Edit		Delete
Name			Value	Type	C I.	3D
bufferGreenParts			(?)	object	*	
bufferReadyParts			(?)	object	*	
successor			(?)	object	*	

Save your changes by clicking OK. You can now assign the two buffers (buffer-GreenParts → P1, bufferReadyParts → P2) to Machine1 and F1 as successor. To do this, click the tab USER-DEFINED ATTRIBUTE in the object Machine1, double-click the relevant line and select the buffer or L1.

Importer		Failure Importer		User-defined Attributes
	New		Edit	Delete

Name		Value	Type	C	I.	3D
bufferGreenParts		P1	object			
bufferReadyParts		P2	object			
successor		L1	object			

This makes programming the method easier for many applications:

- machine1 will be replaced by ?
- p1 will be replaced by ?.bufferGreenParts
- p2 will be replaced by ?.bufferReadyParts
- L1 will be replaced by ?.successor

Specific instructions for a machine are written into an "if then else …" statement.

Method load:

```
is
do
   -- load the part into the palette
  @.move(?.bufferReadyParts.cont);
  if ?.bufferReadyParts.cont.full then
    ?.bufferReadyParts.cont.move(?.successor);
    ?.bufferGreenParts.cont.move(
    ?.bufferReadyParts);
    -- create a new palette
   if ? = machine1 then
        createPallet(.MUs.pallet,.MUs.part,start);
   end;
  else
   -- already parts on p1
   ?.bufferGreenParts.cont.cont.move(?);
  end;
end;
```

Likewise, you can convert the method loadFirstPart. Define an attribute "ma-chine" in the class of the PlaceBuffer (type object). Set Machine1 as the value for the attribute machine in the buffer P1. The method should look like this:

```
is
do
  -- move the first part to the machine
  @.cont.move(?.machine);
end;
```

Now you can easily extend the simulation without the need to write new methods. You have only to assign the methods to the sensors and set the attributes of the objects.

Example 76: Saw

You are to simulate the following process. A block with an edge length of 40 cm is to be sawed into 16 parts. Ten parts each will then be packed into a box. Ten boxes are packed in a carton. Between the saw and the individual packing stations, lines with a length of 5 meters will be set up. Create the folder "Saw" below models. Duplicate all required classes in this folder.

Create the following Frame:

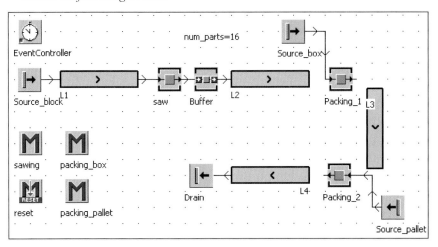

Settings: L1, L2, L3, and L4 length 5 meters, speed 1 m/s; saw: processing time 10 seconds, packing_1 and packing_2 two seconds processing time, drain 0 sec. processing time. Create two entities (block, part) and two containers (box, palette). Block: 0.4 meter length, part 0.1 meter length. Change the icon of the part to a size of 7 x 7 pixels. Box: container, capacity 10 parts, palette: container, capacity 16 boxes. Arrange your sources so that they produce the correct type of MU (e.g., source_box).

| Operating mode: | ☑ Blocking | ☐ |
| Time of creation: | Interval Adjustable ☑ ☐ | |

                                                        DDD:HH:MM:SS.XXXX

Interval:	Const ☑	0:06.25	⊟
Start:	Const ☑	0	☐
Stop:	Const ☑	0	☐

| MU selection: | Constant ☑ ☐ | |
| MU: | .Models.saw.box | [...] ⊟ |

Interval palette: 1:40, interval block: 0:10

Method sawing (exit control front of SingleProc saw): The method must destroy the block and create a certain number of parts. Creating the parts works best with a buffer object. The processing time of the sawing can be considered as processing time of the SingleProc. To make the simulation more flexible, define the number of parts outside of the method (e.g., in a global variable in the example: num_parts).

Method sawing (exit control front saw):

```
is
   i:integer;
do
   -- destroy block
   @.delete;
   -- create num_parts
   for i:=1 to num_parts loop
      .Models.saw.part.create(buffer);
   next
end;
```

At the end of line L2, the parts are to be packed into boxes. If a box is placed on the station Packing_1, the incoming part is transferred to the box. If the box is full, it will be transferred to line L3.

Method packing_box (exit control L2):

```
is
   box:object;
do
   -- wait for a box
   waituntil packing_1.occupied prio 1;
   box:=packing_1.cont;
   -- pack parts into the box
```

```
@.move(box);
if box.full then
   box.move(L3);
end;
end;
```

Similarly, you need to program the method packing_palette. To ensure a smooth start of the simulation, destroy the MUs when resetting the simulation.

Note:
The task can also be solved with dismantle and assembly stations.

Example 77: Kanban Control

Simulations regarding the flow of materials have a fixed direction. The sources produce parts according to a fixed schedule (e.g., batch). The parts move from the source to the drain and trigger the production at the machines (push-control). Many companies (especially in Japan) use the opposite control concept. There, the succeeding stations trigger the production of the preceding stations. Trigger and main information carrier within this system is the Kanban card. We will simulate a Kanban container system. Create the following Frame. Also create the methods:

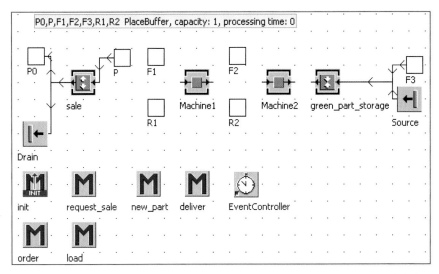

Settings: Machine1 and Machine2 processing time: 1 minute, 100% availability; sale: DismantleStation, successor number 1: P0 (for containers), successor 2: Drain 1 minute processing time. Create an entity .mus.part. The source produces .mus.part, interval 1 minute. Select the following settings in the DismantleStation sale:

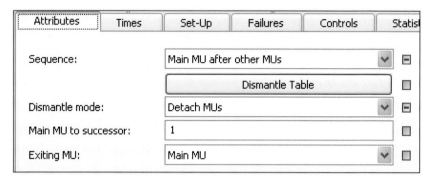

Enter the following information into the dismantle table:

	MU	Number	Successor
1	.MUs.part	20	2

The assembly station green_part_storage loads 20 parts produced by the source onto a container that is transferred from F3. First, connect F3 with green_part_ storage, then to the source. Select the following settings in the station green_ part_storage:

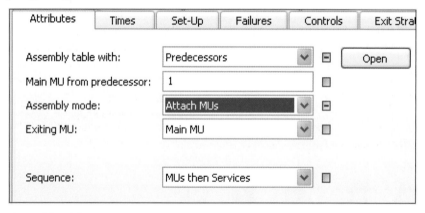

Assembly table:

	Predecessor	Number
1	2	20

The production flow in this model should be like this: A container is located on the station sale and is unloaded gradually. If the container is empty, it will be trans- ferred to the station F1 (finished part place of Machine1). The arrival of the con- tainer triggers the order of the unfinished parts. This takes place by transferring an empty container from the place R1 (unfinished part place of Machine1) to the place F2 (finished part place of Machine2). Machine2 sends a container from R2 to F3 and in this way triggers the delivery of the unfinished parts. The station green_part_storage loads the container with parts and sends it back to R2. The arrival of the container initiates the production at Machine2. If the finished part container of Machine2 (F2) is full, it will be transferred to the unfinished part place of Machine1 (R1). The finished parts of Machine1 will be transferred to the station sale (P1). Trigger and main control tool of the production are the kanban containers. They contain all information required for controlling the production. In your simulation model you can accomplish this with user-defined attributes. Create the container Kanban_container (capacity: 20 parts) in the class library. Create the following user-defined attributes:

Name	Value	Type	C	I.	3D
number	20	integer	*		
part	*.MUs.part	object	*		
processing_time	1:00.0000	time	*		
target_empty	(?)	object	*		
target_full	(?)	object	*		
workplace	(?)	object	*		

Step 1:
Create a filled container on P and empty containers on R1 and R2. Set the necessary information in the kanban containers. A kanban system represents a system of self-regulating control loops. For the present simulation, this means that a container shuttles between two places. The container, which is located on the station sale, shuttles between Machine1 and sale. Empty containers will be transported to F1, full containers always to P. The unfinished parts container of Machine1 (place R1) shuttles between Machine1 and Machine2 (place F2). In other words, if the container is full, it is transported to R1, if it is empty always to F2, etc. For each container, this information has been stored in user-defined attributes. The required init method should look like this:

```
is
   container:object;
do
    deleteMovables;
   -- initialisieren
   -- create a container at sale, load it
   -- target_empty: F1
   -- target_full: sale
   -- workplace: machine1
   container:=.MUs.kanban_container.create(p);
   container.target_empty:=F1;
   container.target_full:=P;
   container.workplace:=machine1;
```

```
while not container.full loop
    .MUs.part.create(container);
end;
-- create kanban_container at r1 and r2
container:=.MUs.kanban_container.create(r1);
container.target_empty:=F2;
container.target_full:=R1;
container.workplace:=machine2;
container:=.MUs.kanban_container.create(r2);
container.target_empty:=F3;
container.target_full:=R2;
container.workplace:=green_part_storage;
end;
```

Step 2:
If the container on the station sale is empty, it orders new parts from Machine1 by sending the container to the station F1. Method request_sale, exit control front P0:

```
is
do
    -- empty container
    -- move to target_empty
    @.move(@.target_empty);
    -- request finished parts
end;
```

Step 3:
On arrival, a container on the finished part station, the machine has to send an unfinished part container as a request to the preceding workplace. To enable more convenient programming of this function, define two user-defined attributes in the class of the SingleProc in the class library:

Name		Value	Type	C	I.	3D
bufferGreenParts		(?)	object	*		
bufferReadyParts		(?)	object	*		

Type the machines into the respective buffers, for example Machine1:

Name		Value	Type	C	I.	3D
bufferGreenParts		R1	object			
bufferReadyParts		F1	object			

Method order, exit control front F1 and F2:

```
is
    container:object;
do
    -- get a reference to the unfinished parts container
```

```
container:=@.workplace.bufferGreenParts.cont;
-- send unfinished parts container
container.move(container.target_empty);
end;
```

Step 4:
After loading the container with unfinished parts, the container has to be transferred to the first unfinished part place (R2).Method deliver exit control front green_part_storage

```
is
do
   @.move(@.target_full);
end;
```

Step 5:
After the arrival of the unfinished parts in the unfinished parts buffer, the first part is transferred to the machine. Create the user-defined attribute machine in the buffer class in the class library. Set the attribute machine of R2 to Machine2 and of R1 to Machine1.

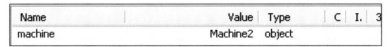

Name	Value	Type	C	I.	3
machine	Machine2	object			

Method load exit control front R1 and R2:

```
is
do
   if @.occupied then
      @.cont.move(?.machine);
   end;
end;
```

Step 6:
After completing processing of the parts on the machine, the machine transfers the part to the container, which is located on the finished part place. If the container is full, then it will be transferred to the station target_full. If the container is not yet full, a new part is loaded onto the machine. Method new_part exit control front Machine1 and Machine2:

```
is
do
   -- load part into the finished part container
   @.move(?.bufferReadyParts.cont);
   if ?.bufferReadyParts.cont.full then
      -- move container
      ?.bufferReadyParts.cont.move(
      ?.bufferReadyParts.cont.target_full);
```

```
    else
    -- load next part
      ?.bufferGreenParts.cont.cont.move(?);
    end;
  end;
```

The simulation now works according to the just-in-time principle.

7.5   The Transporter

7.5.1   Basic Behavior

The Transporter moves on the track with a set speed forward or in reverse. Using the length of the track and the speed of the transporter, the time the Transporter spends on the track is calculated. At the exit, the track transfers the transporter to a successor. Transporter cannot pass each other on a track. If a faster transporter moves up close to a slower one, then it automatically adjusts its speed to the slower transporter. When the obstacle is no longer located in front of the Transporter, the Transporter accelerates to its previous speed. Transporters can have two types of load area:

- Matrix loading space
- Length-oriented loading space

7.5.2   Attributes of the Transporter

Create the object forklift (duplicate a transporter, speed 1 m/s) in the class library. Open the object by double-click it.

Attributes	Failures	Controls	Battery	Product Statistics	Statistics	User-defined

Length:	1.5	m	Booking point:	0.75	m
Speed:	1	m/s	☐ Backwards	☐ Is tractor	
☐ Acceleration			Acceleration:	1	m/s²
			Deceleration:	1	m/s²
☑ Automatic routing		Route weighting:			
Destination:				...	
☐ Matrix load bay					
Load bay length:	1.5		Capacity:	-1	Sensors
Start delay duration:	Const	0			

Length: The length of the transporter must be smaller than the length of the track, if you want to create a transporter on a track. The capacity of the tracks (setting capacity = -1) is calculated as the length of the tracks divided by the length of the transporter.

Speed: Enter the speed with which the transporter moves on the object track. The speed is a positive value (data type real). If you set the speed to 0, the Transporter stops. You can also simulate acceleration and deceleration of the transporter (option **ACCELERATION**).

Backwards: This option activates moving of the transporter in reverse on the track (it also can be called by a method, for example, to drive back the transporter after unloading).

Automatic routing (+destination): If you select this option, then Plant Simulation searches along the connectors for the shortest route to the destination. All objects to the destination must be connected.

Example 78: Automatic Routing

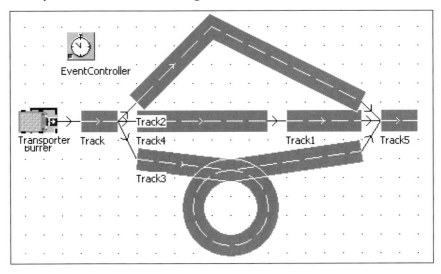

Drag a transporter from the class library to the buffer. Open the dialog of the transporter by double-clicking it. The destination of the transporter is track5.

*Start the simulation and reduce the simulation speed. The transporter finds the
shortest way.*

Matrix load bay: If the option "matrix load bay" is selected, the xy coordinates
then indicate the position of MUs on the load bay. If the box is cleared, Plant
Simulation uses a length-oriented load bay.

Load bay length: Enter the length of the load bay. You can insert sensors and use
the length-oriented load bay like a track, e.g., for representing panel carts, auto-
mobile transporters, loaders, cranes, etc.

Capacity: Enter the number of MUs, which can be located on the transporter,
whole or in part, at any one time. -1 means that no limitations apply. This means
that the loading bay of the transporter is full if all MUs are touching each other.

7.5.3   Routing

If a track has various successors, four different types for routing are available:

- Automatic routing
- Drive control
- Exit control
- Basic behavior

7.5.3.1 Automatic Routing

To use automatic routing, you need to supply the transporter with destination information, and assign information to the track about which destinations are to be
reached on the track (target list). Plant Simulation searches the target lists of the
successors for the destination of the track. Then Plant Simulation transfers the
transporter to the first track whose target list contains the destination.

Example 79: Automatic Routing

*A source randomly produces three parts. A transporter loads the part and trans-
ports it to the relevant machine. Then the transporter drives back to the source.
Each machine can only process one kind of part. A special track leads to each
machine. Create the following Frame:*

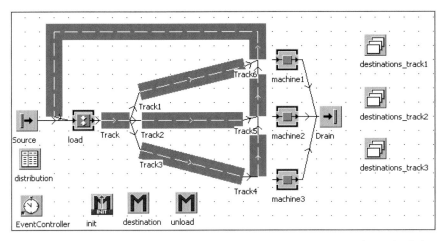

Duplicate the entity three times. Name the parts Part1, Part2, and Part3. Color the parts differently.

Go to the source. Enter an interval of 2 minutes. Select MU Selection random. Enter the table distribution into the text box Table.

MU selection:	Random	▾	⊟	Stream:	1
Table:	distribution	...	⊟	☑ Format table	

Plant Simulation formats the table distribution. Open the table. Drag the parts from the class library to the table (this will enter the absolute path into the table). You can also enter the absolute path yourself. Next to the addresses of the parts, type in the distribution of the parts in relation to the total amount.

	object 1	real 2	string 3
string	MU	Frequencies	Name
1	.MUs.part1	0.10	
2	.MUs.part2	0.20	
3	.MUs.part3	0.70	

The source now produces part1, part2, and part3 in a random sequence.
An assembly station then loads the transporter. First, connect track6 with the assembly station, then with the source. Insert track6 so that the exit is located close to the assembly station. Select the following settings in the assembly station.

One part from the predecessor 2 is to be mounted.

	Predecessor	Number
1	2	1

The processing time of the assembly station is 10 seconds. The init-method inserts the transporter close to the exit of track6.

Method init:

```
is
do
    deleteMovables;
    .MUs.Transporter.create(track6,15);
end;
```

Determine the destination of the transporter depending on the name of the part. Set the value of the attribute destination of the transporter with a method. The output sensor (rear) of the assembly station is to trigger the method.

The method destination sets the attribute depending on the MU names.

```
is
do
    -- @ denotes the transporter
    -- @.cont is the part on the transporter
    if @.cont.name="part1" then
      @.destination:=machine1;
    elseif @.cont.name="part2" then
      @.destination:=machine2;
    elseif @.cont.name="part3" then
      @.zielort:=machine3;
    end;
end;
```

Create and assign the destination list of the tracks. For creating the destination lists, use objects of type CardFile.

*The required data type is object. First, turn off inheritance (**FORMAT – INHERIT FORMAT**).*

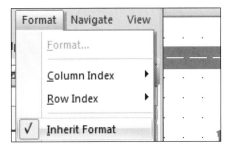

Then click the list header (gray, string) with the right mouse button. Select **FORMAT.** *Select the data type object on the tab data type.*

Now enter the objects, which can be reached via the track. You can also enter the destinations by dragging the objects onto the list and dropping them onto the respective line. This inserts the absolute path. Insert Machine1 into the destinations_list1 and so on.

	object 1
1	.Models.routing.Machine1

Enter the destination list on the tab Attributes of the track (forward destination list).

Forward destination list:	destinations_track1	...

Move the parts at the end of the tracks.The parts are loaded onto the machines at the end of the tracks 1, 2, 3. To accomplish this, we use the destination addresses of the transporters. Enter the method into the exit controls of the tracks 1, 2, and 3(rear).

Method unload:

```
is
do
    -- @ is the transporter
    @.cont.move(@.destination);
end;
```

At the end, the transporter drives by itself to Machine3 on track3 to Machine2 on track2, etc., depending on which part is loaded.

Note:
Plant Simulation transfers the transporter onto the first object in whose destination list the destination of the transporter is registered. If the following track is failed, the transporter stops and waits until the failure is removed. While routing, Plant Simulation does not take the status of the tracks into account.

7.5.3.2 Driving Control

At a junction, you can determine the destination of the transporter with SimTalk, for example, depending on the availability and the load of the target station, and transfer the transporter on the correct track.

Example 80: Driving Control

We want to simulate a manipulation robot (e.g., FlexPicker by ABB). The robot can freely transport parts within a restricted area at high speed. You are to simulate the following problem. The robot takes parts from one place and distributes them onto three lines. The lines have an availability of 98% and an MTTR of 25 minutes. The robot itself has an availability of 99% and an MTTR of 30 minutes. The robot can reach an acceleration/deceleration of 100 m/s² and has a maximum speed of 10 m/s. The cycle time is 1.05 seconds (source interval). The speed of the lines is 0.1 m/s. The part has a length of 0.3 m. The robot has a work area with a diameter of 1.2 meters. Set the scaling factor in the Frame to 0.005.

Create the following Frame:

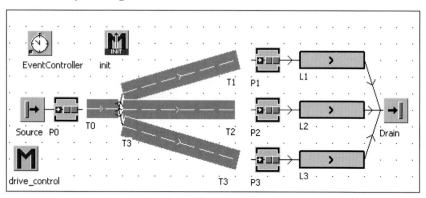

Length of the tracks T0: 0.2 m, W1: 0.75 m, W2: 0.7 m, W3: 0.75 m, processing time P1, P2, P3 3 seconds (to secure a distance between the parts), the capacity of all buffers is one part. The length of the transporter is 0.1 meter (booking point 0). The transporter must drive backwards after being inserted into the frame. Therefore, select Backwards in the dialog of the transporter in the class library. Proceed as follows.

1. Program the Init method. It creates a transporter on the track T0. When creating, a length is passed so that the transporter can trigger a backward exit sensor.

Method init:

```
is
do
    .MUs.Transporter.create(T0,0.1);
end;
```

2. Program the backward exit control drive_control: The transporter waits until a part is located on P0 and loads it. The transporter drives forward until the end of track T0. A single method is to be used for all controls. For that reason, the object and possibly the direction of the transporter will be queried in the method.

Method drive_control:

```
is
do
  if ? = T0 and @.backwards then
  -----------------------------------
  -- T0 exit backwards
    @.stopped:=true;
    waituntil P0.occupied prio 1;
    P0.cont.move(@);
    @.backwards:=false;
    @.stopped:=false;
  -----------------------------------
  end;
end;
```

Assign the method drive_control to the track T0 as the exit control and the backward exit control.

3. Program the Exit control T0: At the end of the track T0, you have to be deciding to which place the transporter is to drive. The transporter waits until P1, P2, or P3 is empty. Starting with the station P1, the method queries whether the place is empty. The transporter will be transferred onto the track to the first empty place. Method drive_control, a new branch in the query if ? = T0 and @.backwards then ...:

```
is
do
  if ? = T0 and @.backwards then
  -----------------------------------
  -- see above
  -----------------------------------
  elseif ?=T0 and @.backwards= false then
    -- T0 exit
    @.stopped:=true;
    waituntil P1.empty or P2.empty or P3.empty prio 1;
    --drive to the empty place
```

```
if P1.empty then
     @.move(T1);
elseif P2.empty then
     @.move(T2);
elseif P3.empty then
     @.move(T3);
end;
  @.stopped:=false;
end;
end;
```

4. Program the Exit control of the tracks T1, T2, and T3: The transporter loads the part into the buffer. After this, the transporter moves backwards to load a new part. To simplify matters, define the attribute buffer (type object) in the class track in the class library and assign the buffer P1 to the track T1, etc.

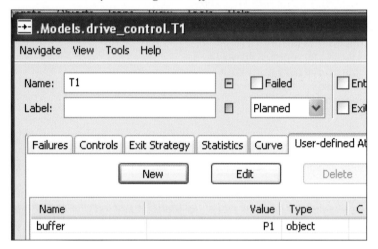

One control only is required for unloading. Therefore, you can program it as an else-block in the query of the objects.

```
is
do
  if ? = T0 and @.backwards then
  -----------------------------------
  -- T0 backwards exit
  -- see above
  -----------------------------------
  elseif ?=T0 and @.backwards= false then
    -- T0 exit
    -- see above
  else
    --unload onto buffer
```

```
    @.stopped:=true;
    @.cont.move(?.buffer);
    @.backwards:=true;
    @.stopped:=false;
  end;
end;
```

7.5.4 Methods and Attributes of the Transporter

7.5.4.1 Creating a Transporter

You can use the method create for creating transporters.

Syntax:

```
<object>.create(target object) or
<object>.create(target object, length)
```

On length-oriented objects, you can determine the initial position at which the transporter will be inserted on the target object.

7.5.4.2 Unloading a Transporter

Unloading of transporters is accomplished analogous to unloading containers, for example, initiated by an exit control of the tracks.

Example 81: Unloading a Transporter

The content of the transporter will be transferred to the machine M2.

```
is
do
  @.cont.move(M2);
end;
```

Explanation: @ denotes the transporter in this case. You can access the part using the method cont of the transporter (@.cont). The method cont returns a reference to the part. You can then transfer the part to the machine M2 with ...move(M2).

7.5.4.3 Driving Forward and Backward

Transporters often shuttle between objects. You need to change direction, so that the transporter can move in the opposite direction of the connectors.

Syntax:

```
@.backwards:=true/false;
```

The attribute backwards returns true, if the direction of the transporter is backward, and false, if the transporter is moving forward. You can set and get the value of the attribute backwards.

7.5.4.4 Stopping and Continuing

To stop and to continue after a certain time is the normal behavior of the transporter. While the transporter waits, you can, e.g., load and unload the transporter or recharge its battery. In SimTalk, you use the attribute stopped to stop the transporter and make it continue on its way.
Syntax:

```
@.stopped:=true;  --stop the transporter
@.stopped:=false;  --the transporter drives again
```

Another possibility to stop the vehicle is to set the speed to 0. The vehicle then slows down with the set acceleration and stops. You can start the vehicle again by setting the speed to its original value. In this way, you can take into account acceleration and slow down in the simulation. To demonstrate these two options, you can use the following example:

Example 82: Stopping Transporters

Create the following frame:

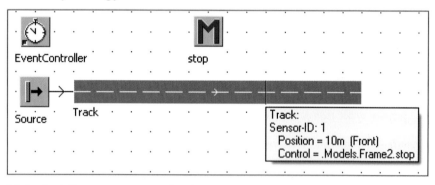

Make the following settings in the class transporter in the class library:

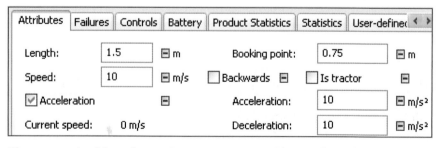

The source should produce only one transporter. This works with the following setting:

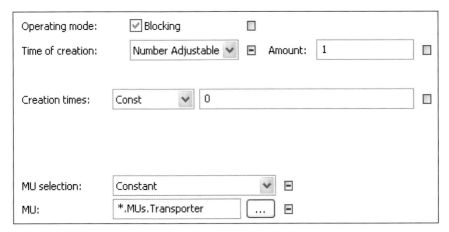

The transporter should stop after 10 meters.

Variant 1: *You insert a sensor in the holding position and trigger at this position a method which stops the vehicle. The slow down is not taken into account. The method stop should look as follows:*

```
(SensorID : integer)
is
do
    @.stopped:=true;
end;
```

If you set the attribute stopped back to false, the transporter moves again.

Variant 2: *The transporter slows down and stops at 10 meters. For this variant, you need a second sensor (approx. 5.1 meters), on wich you start the slow down. The method has in the second variant the following content:*

```
(SensorID : integer)
is
do
    @.speed:=0;
end;
```

The transporter starts again, if you set the speed to a value greater than 0.

7.5.4.5 Drive after a Certain Time

Often the transporter stands for a while before it starts again, for example, for loading and unloading. There are different ways for modeling the standing times of the transporter. Basically, you need to start the transporter with the same manner with which you have stopped it (either with the attribute stopped or speed). Even if the transporter runs at the end of the track and stops by itself, you need to set the attribute stopped: = true to be able to start it later again (with change in direction) without problems.

You can pause the transporter using the method <path>.startPause(<integer>). After <integer> seconds ends the pause. If the transporter is paused, it stops and drives again, if the pause ends.

Example: Transporter Starts after a Certain Time

The transporter from the example above should start again after 10 seconds. The 10 seconds should be taken into account with the method startPause.

Variant 1: transporter stopped with stopped:=true

```
(SensorID : integer)
is
do
    @.stopped:=true;
-- Start again after 10 seconds
    @.startPause(10);
    @.stopped:=false;
end;
```

Variant 2: transporter stopped with speed:=0
It is easiest to use a second sensor to start (and unloading) the transporter again (e.g., SensorID 1 stop, SensorID 2 go). You should make sure that the vehicle, the second sensor also triggers and not stops before. The method could look as follows:

```
(SensorID : integer)
is
do
    if sensorID = 1 then
            @.speed:=0;
    elseif sensorID=2 then
            --start after 10 seconds
            @.startPause(10);
            @.speed:=10;
    end;
end;
```

A second method to start the transporter after a certain time is to use the time of the event controller in combination with a waituntil statement. Therefore, you take at a certain time the simulation time of the event controller and add a certain amount of time to it (e.g., 10 seconds). The simulation time (simTime) of the event controller is observable. Then you can interrupt the processing of the method until your set time is reached. An appropriate method could look as follows:

```
(SensorID : integer)
is
```

```
    simulTime:time;
do
    if sensorID = 1 then
            @.speed:=0;
    elseif sensorID=2 then
            simulTime:=
            eventController.simTime+num_to_time(10);
            --start again after 10 seconds
            waituntil eventController.simTime >=
            simulTime prio 1;
            @.speed:=10;
    end;
end;
```

Note: If no further event is there to process, the event controller stops the simulation. You may have to make a small bypass, so the simulation continues running. In the example above, for example, the following extension reaches (Source1 and Drain).

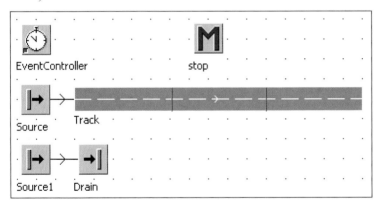

Set the Source1 so that each second one part is produced. This will always generate new events and the simulation continues.

7.5.4.6 Start Delay Duration

A transporter stops automatically when it collides with a standing transporter on the same track. Starts the first transporter again, then all collided transporters automatically start again. To model this behavior more realistic, you can use the attribute start delay duration (e.g., 0.5 seconds). The following transporter will start with a lag of 0.5 seconds, after start of the transporter in front. You can set the start delay duration on the dialog of the transporter (class library):

Example 83: Start Delay Duration, Crossroads

You are to simulate a simple crossroads. The crossroads is regulated by traffic lights. In this example, the transporter decides directly at the crossroads, if it stops or goes on (without slow down). All transporters behind it drive against it. Insert a traffic_light in the class library (duplicate a class SingleProc and rename it). Create two icons in the class traffic_light (icon1: green, icon 2: red). Insert in the class traffic_light a user-defined attribute "go" (data type boolean). Create a new frame. Set the scaling factor to 0.25 (Frame window – Tools – Scaling factor). Set up the following frame:

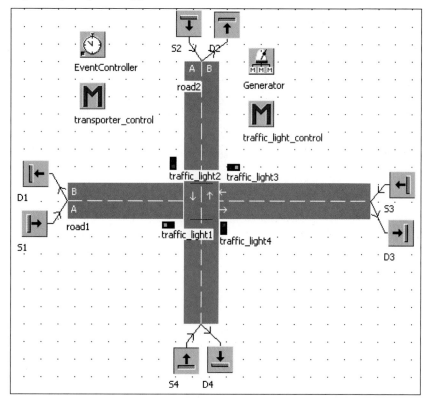

In this example, we also show how the TwoLaneTrack works.

Settings: The sources S1, S2, and S3 create each transporter. The interval of the creation should be randomly distributed. Make the following setting in the source S1:

| Operating mode: | ☐ Blocking | ⊟ |
| Time of creation: | Interval Adjustable ⌄   ☐ | |

		Stream, Start, Stop
Interval:	Uniform ⌄	1,0:05,0:30	⊟
Start:	Const ⌄	0	☐
Stop:	Const ⌄	0	☐

| MU selection: | Constant ⌄   ☐ | |
| MU: | *.MUs.Transporter   [...]   ⊟ | |

Make this setting also for S2, S3, and S4. Set the stream (first number in field interval9 for each source to another value). The transporters move at a speed of 10 meters per second and accelerate with 10 m/s². To ensure that the transporter can pass each other, set in the ways a track pitch of 4 meters.

| Track pitch: | 4 | ⊟   m | Destination lis |
| | Right-hand Traffic ⌄   ☐ | | |

Create at the crossroads sensors on the lanes. You can specify for each lane its own sensors. You must uncheck for the other lane the checkboxes for the front and rear. For every track, you need to specify two sensors. One sensor in lane A and one sensor in lane B. All sensors will trigger the method transporter_control. In case of road1, it could look as follows:

ID	Position	Front	Rear	Path
1	35m / 55m	A		transporter_control
2	45m / 45m	B		transporter_control

Initialize the simulation, so that two traffic lights show the icon 2 (red) (right mouse button – next icon) and the attribute go is false; the other two traffic lights show the green icon and the attribute go has the value true.

Traffic light control

For the traffic light control, we use in this example a method (traffic_light_control) and a generator. The method switches the lights and the generator repeatedly calls the method with an interval of 1:30 minutes.
Method traffic_light_control:

```
is
do
    --switchs the traffic light
    -- icon1 green, icon2 red
    if traffic_light1.go then
            traffic_light1.go:=false;
            traffic_light1.CurrIconNo:=2;
            traffic_light3.go:=false;
            traffic_light3.CurrIconNo:=2;

            traffic_light2.go:=true;
            traffic_light2.CurrIconNo:=1;
            traffic_light4.go:=true;
            traffic_light4.CurrIconNo:=1;
    else
            traffic_light1.go:=true;
            traffic_light1.CurrIconNo:=1;
            traffic_light3.go:=true;
            traffic_light3.CurrIconNo:=1;

            traffic_light2.go:=false;
            traffic_light2.CurrIconNo:=2;
            traffic_light4.go:=false;
            traffic_light4.CurrIconNo:=2;
    end;
end;
```

You can test the method by starting this repeatedly. The lights should "switch". The method is called by the generator. Open the generator by double-click and make the following settings on the tab times:

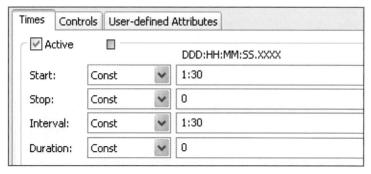

Enter the method traffic_light_control in the field Interval on the tab Controls (more in chapter information flow objects).

```
┌──────────┬──────────────┬──────────────────────────┐
│  Times   │  Controls    │ User-defined Attributes    │
├──────────┴──────────────┴────────────────────────────┐
│                                                        │
│   Interval:   ┌────────────────────────────┐ ┌─────┐ □ │
│               │ traffic_light_control       │ │ ... │ ⊟ │
│               └────────────────────────────┘ └─────┘   │
│   Duration:   ┌────────────────────────────┐ ┌─────┐ □ │
│               │                             │ │ ... │ □ │
│               └────────────────────────────┘ └─────┘   │
│                                                        │
└────────────────────────────────────────────────────────┘
```

The transporter should stop when the relevant traffic light has the value go=false and wait until go=true. Then the transporter should start again. If the lane A has sensor number 1 and the lane B has sensor number 2, the method transporter_ control should look as follows:

```
(sensorID : integer)
is
do
    if ?=road1 and sensorID=1 then
        if traffic_light1.go=false then
            --traffic_light is red - stop
            @.stopped:=true;
            waituntil traffic_light1.go prio 1;
            @.stopped:=false;
        end;
    elseif ?=road1 and sensorID=2 then
        if traffic_light3.go=false then
            @.stopped:=true;
            waituntil traffic_light3.go prio 1;
            @.stopped:=false;
        end;
    elseif ?=road2 and sensorID=1 then
        if traffic_light2.go=false then
            @.stopped:=true;
            waituntil traffic_light2.go prio 1;
            @.stopped:=false;
        end;
    elseif ?=road2 and sensorID=2 then
        if traffic_light4.go=false then
            @.stopped:=true;
            waituntil traffic_light4.go prio 1;
            @.stopped:=false;
        end;
    end;
end;
```

Change the start delay duration in the class transporter (class library) to 0.5 seconds and watch what happens.

Note:
If the transporter does not stop exactly on the line of the sensor, then the reference point of the vehicle is in the wrong position. The reference point is in the default icon in the middle of the symbol. If the transporter is to stop exactly on the line, then you need to set the reference point to the right edge of the icon.

7.5.4.7 Important Methods and Attributes of the Transporter

Method/attribute	Description

`<MU-path>.startPause;`
`<MU-path>.startPause`
`(<time>);`
	The method `startPause` immediately pauses the Transporter and sets the attribute pause to the value true. When a parameter is passed (integer greater than 0), it determines after which time (in seconds) the transporter changed back to the non-paused state.

`<MU-path>.startPauseIn` `(<time>)`	Pauses the transporter after the period defined in `<time>` has passed.
`<MU-path>.collided;`	Collided returns true if the transporter is collided with another transporter.
`<MU-path>.XDim;` `<MU-path>.YDim;`	Sets/gets the dimension of the matrix load bay
`<MU-path>.speed;`	Specifies the speed with which the transporter moves on the track. The speed must be equal to or greater than 0. If you set the speed to 0, then the transporter stops.
`<MU-path>.destination;`	Sets/gets the destination of the transporter

The Transporter also provides a number of methods and properties, which deal with the battery operation and related problems.

Example 84: Portal Loader Parallel Processing

You are to simulate a portal loader which loads two machines. The machines simultaneously process the same kind of part. The loader picks up parts at a place at the beginning of the track parts (capacity one part) and distributes them to the machines. If both machines are loaded and working, the loader waits empty between the two machines. When the first machine has finished processing, the loader drives to the machine and unloads it. A time of 5 seconds for the handling by the loader is considered (the movement in the z-axis is not simulated). Create the following Frame:

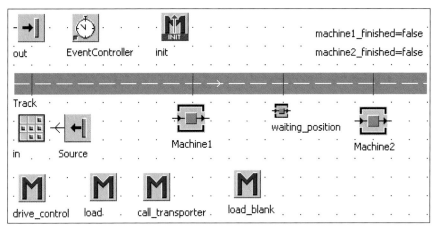

Settings: The source produces parts at an interval of 3:30 minutes. The processing time of Machine1 is 7:50 minutes, of Machine2 7:40 minutes. Both machines have an availability of 90% and an MTTR of 45 minutes. The track has a length of 25 meters; the transporter has a length of 1.5 meters, a speed of 1 m/s and a capacity of one part. The capacity of the object in is one part. Insert the following sensors into the track (as the drive_control):

ID	Position	Front	Rear	Path
1	1m		x	drive_control
2	10m	x		drive_control
3	15m	x		drive_control
4	20m	x		drive_control

The global variables machine1_finished and machine2_finished have the data type boolean and the initial value false.

1. Insert the transporter: The init method creates the transporter. Prior to that, all MUs will be destroyed.

Method init:

```
is
do
    deleteMovables;
    .MUs.Transporter.create(track,12);
    track.cont.destination:=in;
    track.cont.backwards:=true;
end;
```

2. Program the driving control: The transporter will be addressed for each trip. For this, the attribute destination is used. Destinations are assigned to certain sensor IDs. Once the transporter arrives at the destination (target sensor ID), the transporter stops. The method drive_control needs a parameter for the sensor_ID. First, the transporter is to stop at the sensor_ID 1.

Method drive_control:

```
(sensorID : integer)
is
do
    if sensorID=1 and (@.destination=in
    or @.destination=out) then
        @.stopped:=true;
    end;
end;
```

3. Program the method for loading the unfinished part: The method load_blank uses the transporter as the parameter and should have the following functionality: The method waits until the place "in" is occupied. If Machine1 or Machine2 is empty, the transporter loads the part and drives to the empty machine. If both machines are occupied, the transporter drives to the waiting position. The handling time is taken into account by pausing the transporter.

Method load_blank:

```
(transporter:object)
is
do
    --search an empty and operational machine
    --wait for an unfinished part
    waituntil in.occupied prio 1;
    if machine1.empty and machine1.operational then
            in.cont.move(transporter);
            transporter.destination:=machine1;
    elseif machine2.empty and machine2.operational
    then
            in.cont.move(transporter);
            transporter.destination:=machine2;
    else
            transporter.destination:=waiting_position;

    end;
    --drive forward
    transporter.backwards:=false;
    -- 5 seconds handling time
    transporter.startPause(5);
    transporter.stopped:=false;
end;
```

Call within the method drive_control: If a loaded transporter arrives at sensor 1, first the part is transferred to the drain "out", then the method load_blank is called.

Method drive_control:

```
(sensorID : integer)
is
do
    if sensorID=1 and (@.destination=in or
        @.destination=out) then
            @.stopped:=true;
            if @.empty then
                    load_blank(@);
            else
                    -- move parts to the drain
                    @.cont.move(out);
                    --load new parts
                    load_blank(@);
            end;
    end;
end;
```

4. Program the method load: The loaded transporter is to stop at the machine and load the part onto the machine (if the machine is operational). Thereafter, the transporter moves to the waiting position. If the transporter is empty, the transporter unloads the machine and drives to the drain "out". The method load requires three parameters: machine, transporter, and the direction to the waiting position (backwards true/false). The method load could look like this:

```
(transporter:object;machine:object;
directionWaitingPosition:boolean)
is
do
    transporter.stopped:=true;
    if transporter.occupied then
    -- transporter is loaded
            waituntil machine.operational prio 1;
            transporter.cont.move(machine);
            transporter.destination:=waiting_position;
            transporter.backwards:=
                directionWaitingPosition;
    else
    -- transporter is empty
            machine.cont.move(transporter);
            transporter.destination:=out;
            transporter.backwards:=true;
    end;
```

```
-- start after 5 seconds
transporter.startPause(5);
transporter.stopped:=false;
end;
```

5. Program the method call_transporter: If the machines are ready, they must send a signal. The class SingleProc provides the method ready that returns true if the station has finished processing parts. This value, however, is not observable. One solution would be the following: If the machine is ready, the processed part then triggers a control that sets a global variable to true (e.g., machine1_finished). Global variables are observable and an appropriate action can be triggered. Within the frame, the ready variables consist of the machine name and "_finished". A universal method for registering the finished machines could look like this:

Method call_transporter:

```
is
do
  -- ? object, that calls
  str_to_obj(?.name+"_finished"):=true;
end;
```

str_to_object converts a string (object name) to an object reference. Assign the method call_transporter to the exit control (front) of Machine1 and Machine2.

6. Complete the method drive_control: Within the drive_control, the method load must be called at the positions of Machine1 and Machine2. A control for the waiting position is established at the position of the waiting position: If Machine1 or Machine2 is operational and empty, the transporter drives into the station and delivers a new part. Otherwise, the transporter waits until Machine1 or Machine2 is ready. The transporter might change its direction, set a new destination, and set the finished variable of the machine to false. Finally, the transporter drives to the machine. The completed method drive_control should look as follows:

```
(sensorID : integer)
is
do
  if sensorID=1 and (@.destination=in or
  @.destination=out)
  then
    @.stopped:=true;
    if @.empty then
      load_blank(@);
    else
      -- move parts to the drain
      @.cont.move(out);
      -- load new parts
```

```
      load_blank(@);
    end;
  elseif sensorID=2 and @.destination = machine1
  then
    load(@,machine1,false);
  elseif sensorID=4 and @.destination = machine2
  then
    load(@,machine2,true);
  elseif sensorID=3 and @.destination =
         waiting_position then
    stopped:=true;
    --new part if one machine is empty and
    --operational
    if (machine1.empty and machine1.operational)  or
       (machine2.empty and machine2.operational)
    then
      @.destination:=in;
      @.backwards:=true;
    else
      -- wait until one machine has finished
      waituntil machine1_finished or
      machine2_finished
      prio 1;
      if machine1_finished then
        @.destination:=machine1;
        @.backwards:=true;
        machine1_finished:=false;
      elseif machine2_finished  then
        @.destination:=machine2;
        @.backwards:=false;
        machine2_finished:=false;
      end;
    end;
    @.stopped:=false;
  end;
end;
```

8   Information Flow Objects

The information objects are used for managing information and data. In addition to the global variable and the method, the following objects are information flow objects:

- Lists and tables
- Trigger and Generator
- AttributeExplorer
- Objects for data exchange

8.1   The List Editor

Enter and select settings and entries in lists and tables in the list editor. Duplicate the object TableFile in the class library, and open the duplicate by double-clicking it. Each column has a data type; each cell has a unique address.

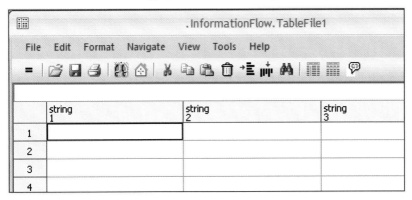

If you want to type data into a table or list in a frame, you have to turn off inheritance: FORMAT – INHERIT FORMAT

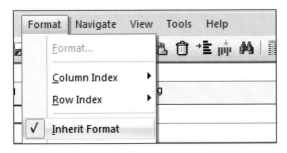

S. Bangsow: Manufacturing Simulation with Plant Simulation, Simtalk, pp. 183–221, 2010.
© Springer Berlin Heidelberg 2010

For setting the data types of each column, it is best to use the context menu (right mouse button on the column header) **FORMAT** …

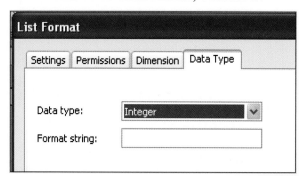

Select the data type of the column here. The format string restricts interactive input (validity of entries). -9, for example, means that numbers with a maximum nine digits can be entered, negative numbers are allowed.

Permissions: On the tab permissions you can endow cells with write protection.

Column/ Row Index: You can define your own column and row indices. This is very helpful especially when you are working with tables. By default, the row and column indices are hidden.

You can view the indices by selecting **FORMAT – COLUMN INDEX – ACTIVE** or **FORMAT – ROW INDEX – ACTIVE**

Example Supply List:

	string 0	object 1	integer 2	string 3	table 4
string		MU	Number	Name	Attribute
2					
3					

The row and column index each have the index 0.

8.2   The CardFile

The CardFile is a one-dimensional list with random access to the content via an index. You can use the CardFile like an array, so that you can store and read many values under one name. This object can easily store data with the same data type. When you insert entries, Plant Simulation moves the following entries back one position. You can remove entries (with "[]"). However, you can also read entries without removing the entry from the CardFile (with the command read).

Example 85: Materials List

Insert a CardFile with the name "material" into an empty frame. Turn off inheritance. Enter the following values:

	string 1
1	wood
2	stone
3	steel
4	reed

A method now is to read line 2 (define a method in the same frame). The following commands are required to do so.

```
is
do
    print material.read(2);
end;
```

Run the method (run-run or F5). The console should display "stone". "Gold" should be inserted in line 3. For inserting entries the CardFile provides the method insert (position, value). Change the method as follows:

```
is
do
    material.insert(3,"gold");
end;
```

Run the method with F5!

The most important attributes and methods of the CardFile are

Method/Attribute	Description
`<path>.insert(<integer>, <value>);`	Inserts the value `<value>` at the position `<integer>`. Entries with the same or a higher index will be moved. In QueueFile and StackFile you only pass the value; in the QueueFile, insertion takes place at the last position, and in the StackFile at the first position
`<path>.cutRow(<integer>);`	Removes the entry with the index `<integer>`, all other entries move up. The method returns the re-moved entry. For QueueFile and StackFile, you pass no index. In the QueueFile, the method deletes the first position, in the Stack-File the last.
`<path>.read(<integer>);`	Reads the entry with the index `<integer>` without removing it.
`<path>.append(<value>);`	Appends the passed value to the end of the list

<path>[<integer>]	Returns the value at the position <integer> and deletes the entry
<path>.dim	Returns the number of entries
<path>.empty	Returns true if the list contains no entries
<path>.delete	Deletes the whole list

Example 86: Handling by a Robot

You are to simulate a robot, which loads several machines. The robot takes the parts from a buffer and loads them into two machines. The robot has a swivel range (diameter) of 4 meters. The positions of the machines are located relative to the robot at 90° and 135° clockwise. Create the following Frame:

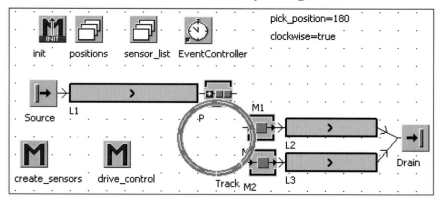

Settings: Source interval 30 seconds, Buffer P processing time: 0 seconds, capacity one part; M1 and M2 each 55 seconds processing time; L1, L2, and L3 each 1 m/s speed.

Creating the Circular Track

For the robot to show the correct behavior, you must create the track as follows. The easiest way is to insert the track from the toolbox. First, click the track button in the toolbox. Then click in the frame. Press Ctrl + Shift. This activates curve mode.

1. Click, counter-clockwise, upward curve	2. Click, coun-terclockwise	3. Click, coun-terclockwise	4. Click, coun-ter-clockwise, click the right mouse button to exit

You must insert a connector from the end of the track to the beginning of the track. If you need a different radius, enter these settings into the dialog EDIT PARA-METER OF CURVE:

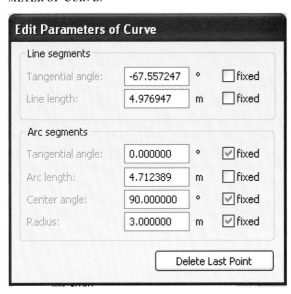

If the option FIXED is not selected, you can set the radius of the arc segment by dragging the mouse. Then, proceed like this:

Design the robot: The robot consists of two parts the robot propper, which moves on the track (in reality Plant Simulation rotates the icon), and a gripper, which moves on the robot (forward and backward). Both are transporters. Duplicate two transporters, and rename them to robot and gripper. Select the following settings in the transporter robot: The robot has a length-oriented load bay with a length of 0.5m:

Attributes	Failures	Controls	Battery	Product Statistics	Statistics	User-defined

Length:	0.1	⊟ m		Booking point:	0.005	⊟ m
Speed:	2	⊟ m/s	☐ Backwards ⊟	☐ Is tractor		⊟
☐ Acceleration		⊟		Acceleration:	1	⊟ m/s²
				Deceleration:	1	⊟ m/s²
☐ Automatic routing		⊟	Route weighting:			⊟
Destination:					...	⊟
☐ Matrix load bay		⊟				
Load bay length:	0.5	⊟ m	Capacity:	-1	⊟	Sensors

Icon operational, pause, failed, waiting: You have to create an icon that can be rotated by Plant Simulation. Therefore, you must set the reference point to the edge of the icon (in this example at the bottom, middle). The track has the setting **CURVE ROTATE MUS** *on the tab curve.*

If the transporter now drives on the track (the reference point is located on the track), Plant Simulation rotates the icon to the position of the transporter on the track. This only works if you insert the track counterclockwise (see above). You must set the length of the transporter on a very small value (e.g., 1 mm), so that the movement matches the position of the transporter on the circle otherwise, e.g., the transporter triggers a control with the front and stops a few angular degrees before the actual position. Try to follow these guidelines for creating an icon for the robot (81 x 81 pixels). The robot is created at the end of the track. The end of the track is located "below", so the icon of the robot must also point downward.

Draw mode	*Animation mode*
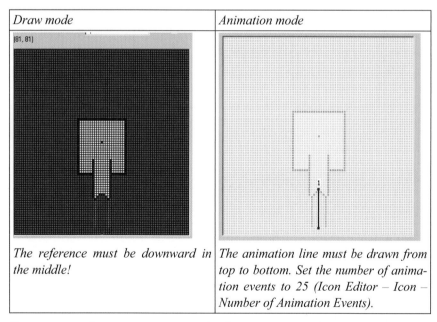	
The reference must be downward in the middle!	*The animation line must be drawn from top to bottom. Set the number of animation events to 25 (Icon Editor – Icon – Number of Animation Events).*

To test your robot, you can insert it by dragging it onto the track. The position should look as follows:

Correct: (track is drawn counterclockwise), the reference point is located at the edge of the icon	Wrong: Path drawn clockwise	Wrong: The reference point is not located at the edge of the icon
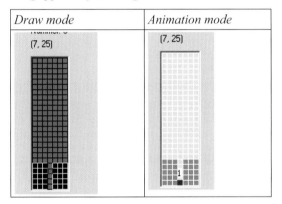		

The gripper may be designed like this:

Draw mode	Animation mode
(7, 25)	(7, 25)

The reference point is located at the lower end of the icon. There also is an animation point for the MU which is to be transported. The orientation and size must be coordinated with the robot.

2. Create sensors on the track: The angles relative to the loading position are known. The loading position in this example is located at 180° (0° is down). The positions of the machines are 90° and 135° clockwise from the loading position. Handling robots especially calculating the sensor positions can be very time-consuming. The sensors are created by a method. The method will delete all old sensors from the track and then create new sensors from a list and attach a predetermined method (drive_control).

Working with Sensors in SimTalk

There are two ways to access the sensors on a track, a conveyor or a transporter via SimTalk:

```
<path>.sensorID(<integer>),
```

via the sensorID (if known) or

```
<path>.sensorNo(<integer>),
```

via an enumeration. You can query the number of sensors with the method

```
<path>.numSensors
```

You create new sensors using the method:

```
<path>.createSensor(<integer>,<string>,<object>,
<boolean1>, <boolean2>)
```

You must indicate the position of the sensor, the type of position ("length" or "relative"), the method to be executed (so that the method is not called by its name, use the method ref ()), and a boolean value for the front and rear control. Sensors will be destroyed using the method

```
<path>.deleteSensor(<integer>)
```

You must pass the ID of the sensor to be deleted. The easiest way to delete all sensors is to delete the first sensor repeatedly, until no sensor is left.

The sensor itself has the following attributes and methods:

Attribute/ method	Description
`<sensor>.position`	returns the position of the sensor, if the position type is "relative" the returned value is a percentage value based on the length of the object, if the position type is length, the returned value is a length position on the object
`<sensor>.front`	sets and gets whether the front control is enabled
`<sensor>.rear`	sets and gets whether the rear control is enabled
`<sensor>.positionType`	gets and sets the position type, possible values are: length or relative

Example: The frame includes two global variables:

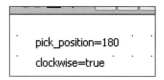

```
pick_position=180
clockwise=true
```

Enter the following values in the table positions (positions of the sensors from the pick position):

	real 1
1	90.00
2	135.00

The method create_sensors is to first delete all sensors, then create the pick position as sensorID 1, then create all sensors from the list positions (at the relevant positions).

Method create_sensors:

```
is
   i:integer;
   id_sens:integer;
   posi:real;
   posi_further:real;
   number:integer;
do
   -- delete all sensors on the track
   number:=track.numSensors;
   for i:=1 to  number loop
     id_sens:=track.sensorNo(1).id;
     track.deleteSensor(id_sens);
   next;
   -- pick_position as first sensor
   posi:=track.length/360 * pick_position;
   --insert the sensor in the track
   id_sens:=track.createSensor(posi,"Length",
           ref(drive_control),true,false);
   for i:=1 to positions.dim loop
     if clockwise then
     -- track is counterclockwise, subtract positions
     -- less than  0 --> subtract the rest from the
     --end of the track
     posi_further:=posi-positions.read(i)*
                 track.length/360;
     if posi_further<0 then
       posi_further:=track.length+posi_further;
     end;
     --create the sensor
     id_sens:=track.createSensor(posi_further,
         "Length", ref(drive_control),true,false);
     end;
   next;
end;
```

3. Create sensors on a transporter, program the control of the gripper: The grip-per should have (here simplified) two positions. One position is at the end of the robot for loading and unloading and the other at the beginning of the robot at which the movement of the robot (here its rotation) is triggered. The control of the gripper can then be quite simple. The robot sets the destination of the gripper and starts the forward movement of the gripper. At the end of the track, the gripper loads the loaded part onto the destination object. If the gripper is empty, then the part is loaded from the destination object. The robot controls the correct timing and the technological sequence. When the gripper has finished its work, it moves back and its rear triggers the movement of the robot as such.

Note:
We recommend to program the control for the gripper in the class library. The method for the robot has to be addressed with its absolute path; otherwise, problems with the instantiation might arise.

Create the method gripper_control in the class library. Then open the transporter robot (0.5 m long) and click the button Sensors:

Insert two sensors, and assign the method gripper_control to the sensors (important, as an absolute path!):

ID	Position	Front	Rear	Path
1	0m		x	.Models.gripper_control
2	0.5m	x		.Models.gripper_control

In its most basic form, the method gripper_control could look like this:

```
(sensorID:integer)
is
   target:object;
do
   if sensorID=2 then
   -- first front
   @.stopped:=true;
      if @.numMU >0 then
         target:=@.destination;
         waituntil target.empty and target.operational
         prio 1;
         @.cont.move(target);
         @.backwards:=true;
         -- pause ?
         @.stopped:=false;
```

```
    else
      -- empty
      -- wait for parts
      target:=@.destination;
      waituntil target.occupied prio 1;
      target.cont.move(@);
      -- start gripper
      @.backwards:=true;
      @.stopped:=false;
    end;
  elseif sensorID = 1 then
    -- turn the robot
    @.stopped:=true;
    ?.stopped:=false;
  end;
end;
```

This gripper control also works very well for controlling grippers in machine portals.

4. Insert the robot and the gripper on the track: The control of the robot on the track is to be triggered by SensorIDs. Add the user-defined attribute targetSensorID (integer) to the robot in the class library, and set the start value to 1:

Name	Value	Type	C
targetSensorID	1	integer	*

The robot first moves to the pick position and waits there until the first part arrives. Robot and gripper are created in the Init method:

```
is
do
    deleteMovables;
    .MUs.robot.create(track);
    .MUs.gripper.create(track.cont,0.4);
    -- stop the gripper
    track.cont.cont.stopped:=true;
end;
```

If everything worked as intended, the robot should be located on its track and the gripper on the robot:

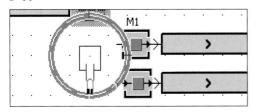

5. Program the method drive_control: When the robot arrives at its target sensor, it stops. The robot waits at sensor 1 until the place p is occupied. Then the robot sets the destination of the gripper to P and starts the gripper (forward). Thereafter, the robot sets the next targetSensorID and the correct value for the moving backwards. The machine should be identified via the SensorID. This is most easily accomplished with a Cardfile (sensor_list). Enter the following data into the sensor_list (data type object):

	object 1
1	p
2	M1
3	M2

Method drive_control:

```
(sensorID:integer)
is
   gripper:object;
   target:object;
do
   gripper:=?.cont.cont;
   if @.targetSensorID=sensorID then
     @.stopped:=true;
     gripper.destination:=sensor_list.read(sensorID);
     -- next target ??
     if gripper.destination= P then
        waituntil (M1.empty or M2.empty) and
        P.occupied  prio 1;
        gripper.backwards:=false;-- drive gripper
        gripper.stopped:=false;
        -- to M1 or M2
        if M1.empty then
           @.targetSensorId:=2;
        elseif M2.empty then
           @.targetSensorId:=3;
        end;
        @.backwards:=true;
     else
        gripper.backwards:=false;
        gripper.stopped:=false;
        @.targetSensorID:=1;-- to pick position
        @.backwards:=false;
     end;
   end;
end;
```

The robot now loads the two machines.

8.3   StackFile and QueueFile

The StackFile and QueueFile are one-dimensional lists which are accessed according to the FIFO (First in first out, queue) or LIFO (Last in first out, stack) principle. New entries will be inserted into the StackFile at the top of the list; the last element inserted is the first entry which will be removed. New entries will be added to the QueueFile at the bottom of the list, the first element will be removed. The main methods for working with stacks and queues are push (element) and pop. With delete, the entire content of a list will be deleted.

Example 87: Queuing

A transporter unloads three machines. It waits in its waiting position (here, 12.5 m), until a driving order arrives. The transporter then drives to the machine and unloads the part. The transporter drives with the part to the end of the track and unloads the part there.

Create the following Frame:

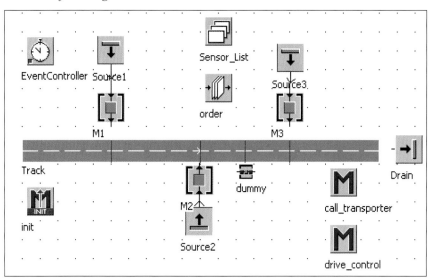

Sensors on the track (length 20 meters):

ID	Position	Front	Rear	Path
1	5m	x		drive_control
2	10m	x		drive_control
3	12.5m	x		drive_control
4	15m	x		drive_control

Settings: Source1, Source2, and Source3 non-blocking, interval 1 minute; M1, M2, and M3 1 minute processing time, availability 50%, 2 minutes MTTR, select different random number streams for the different machines; transporter speed 1 m/s, capacity one part; drain 0 seconds processing time. The transporter has a user-defined attribute targetSensorID (integer). The start value is 3.

Name:	targetSensorID		⊟

Value	Statistics	Communication

Data type:	real	∨
Value:	3	

1. The init-method creates the transporter on the track. The transporter is to stop at the sensorID 3.

The method init looks as follows:

```
is
do
    .MUs.Transporter.create(track,1);
end;
```

The transporter should always stop when the SensorID of the track matches the targetSensorID of the transporter.

Program the method drive_control:

```
(sensorID : integer)
is
do
   if sensorID = @.targetSensorID then
     @.stopped:=true;
   end;
end;
```

2. The machines call the transporter after having processed the parts. In this example, the machine should enter its sensorID into a QueueFile. The assignment of machines to the sensors is entered into a Cardfile (sensor_list). If a sensor is not to be used, enter the object dummy instead (as a wildcard). You cannot leave empty rows in a Cardfile. Also add a dummy object to the frame; otherwise, you will receive an error message. In our example, the sensor list looks as follows:

	object 1
1	M1
2	M2
3	dummy
4	M3

If a part is finished on a machine, it triggers the exit sensor. The machine then has to search the SensorID in the sensor_list and enter the SensorID into the Queue-File order.

Searching in Lists

Searching in lists in Plant Simulation works as follows. Plant Simulation uses an internal cursor for searching. This cursor is set to a hit. With help of the cursor, you can determine the position within the list. At the end of the search, the cursor keeps its old position where it found the entry. Therefore, it is necessary to first set the cursor to the position 1.

Syntax:

```
<path>.setCursor(1);
```

Then you can use the method <list>.find(value) for searching a value. The method find returns true, if the value was found, and false, if the value was not found. If the list includes the queried value, the cursor is set to the corresponding position. In a third step, you need to read the cursor position:

```
position:= <list>.cursor.
```

Example: Program the method call_transporter as the exit control of the machines M1, M2, and M3. Select the data type integer for the QueueFile (order). The method call_transporter could look as follows:

```
is
   sensorID:integer;
do
   -- search sensorID
   -- set the position of the cursor to the beginning
   sensor_list.setCursor(1);
   -- search for the machine
   sensor_list.find(?);
   -- position the cursor
   sensorID:=sensor_list.cursor;
   -- insert into order
   order.push(sensorID);
end;
```

3. The transporter is waiting at the waiting position until an order arrives. Then it drives to the machine. Using the attribute dim, you can determine the number of entries in a list. This attribute is observable; it can be monitored with an observer or a Waituntil statement. The direction can be determined using the sensor IDs. In this case, this is easy because the sensors are not in a mixed order. If you insert a new sensor afterwards (e.g., for a new machine), the sensor IDs get mixed up, which means that a greater sensor IDs do not necessarily mean a greater amount of length of the position.

Sensor Position, Sensor ID, Direction

The direction of a transporter is to be identified. These arguments have to be passed:

- *track*
- *current sensorID*
- *target sensorID*

The return value of the function is of type boolean. Create the method (getDirection) in the current example. Using <track>.SensorID(id) you can access all information that is associated with a sensor. The method SensorID returns an object of type sensor. The attribute position returns the position of the sensor.

Example 88: Determining Sensor Positions

Program the method getDirection:

```
(track:object;sensorFrom:integer;sensorTo:integer)
:boolean
is
  posFrom:real;
  posTo:real;
  backwards:boolean;
do
  posFrom:=track.sensorID(sensorFrom).position;
  posTo:=track.sensorID(sensorTo).position;
  backwards:=(posFrom>posTo);
  return backwards;
end;
```

Extend the method drive_control as follows.

```
(sensorID : integer)
is
do
  if sensorID = @.targetSensorID then
    @.stopped:=true;
    if sensorID= 3 then
      waituntil order.dim > 0 prio 1;
      @.targetSensorID:=order.pop;
      @.backwards:= getDirection(?,3,
                      @.targetSensorID);
      @.stopped:=false;
    end;
  end;
end;
```

4. The Transporter gets an order, drives off, and stops in front of a machine. The transporter has to load the part from the machine and drive forward to the sink. The transporter does not yet know the machine; the method must read the machine from the sensor list. This is accomplished with the method read (id).

Program the method drive_control; in addition to example above

```
(sensorID : integer)
is
do
  if sensorID = @.targetSensorID then
    @.stopped:=true;
    if sensorID= 3 then
      waituntil order.dim > 0 prio 1;
      @.targetSensorID:=order.pop;
      @.backwards:=getDirection(?,3,
          @.targetSensorID);
      @.stopped:=false;
    else
      sensor_list.read(sensorID).cont.move(@);
      @.backwards:=false;
      @.stopped:=false;
    end;
  end;
end;
```

5. Program the method unload for unloading the part onto the drain, and assign it as the exit control of the track. The transporter must unload the part to the drain. Then the transporter drives forward. If an order exists, it must be read and the targetSensorID of the transporter must be set anew. If no order exists, then the targetSensorID is 3.

Program the method unload (exit control track)

```
is
do
  @.stopped:=true;
  @.cont.move(drain);
  if order.dim > 0 then
  -- to machine
    @.targetSensorID:=order.pop;
  else
  -- to waiting position
    @.targetSensorID:=3;
  end;
  @.backwards:=true;
  @.stopped:=false;
end;
```

8.4   The TableFile

8.4.1   Basic Behavior

The TableFile is a two-dimensional list, which allows random access to the entries via their address. TableFiles have many fields of application in simulation projects, e.g.:

- Storage of work plans and production orders
- Collection of statistical information
- Parameterization of models

Note:
For the following example: For the distribution of setup times, depending on the actual part and the new part, you can define the setup time in a matrix.

Example 89: Lot Change
A milling center successively processes different production orders. The parts call for different setup times. Prior to the part moving onto the machine, the setup is to be set anew if necessary. The necessary information is to be centrally stored in a TableFile.

Create the following Frame:

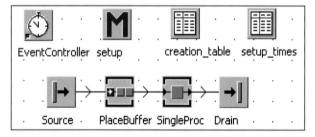

Settings: Source interval 1:30 minutes, blocking; create three entities p1,p2,p3. The source creates the parts in a cycle. Use the TableFile creation_table for the distribution of the parts:

Interval:	Const ∨	1:30	⊟
Start:	Const ∨	0	☐
Stop:	Const ∨	0	☐
MU selection:	Sequence Cyclical ∨	⊟	☐ Generate as batch ☐
Table:	creation_table	... ⊟	☑ Format table ☐

Type in the following lot sizes into the TableFile creation_table:

	object 1	integer 2	string 3
string	MU	Number	Name
1	.MUs.P1	5	
2	.MUs.P2	10	
3	.MUs.P3	5	

PlaceBuffer: capacity 100 parts, no processing time; the TableFile setup_time contains the following information:

	string 0	time 1	time 2
string		setup_time	dismantling_time
1	P1	10:00.0000	5:00.0000
2	P2	13:00.0000	3:00.0000
3	P3	20:00.0000	5:00.0000

The setup time consists of the dismantling time of the old part (already located on the machine) and the setup time for the new part. Setting up from P1 to P2 might take, for example, 5+13 minutes = 18 minutes. The setup time must be assigned to the workstation before the machine starts to set up (automatically every time a MU with a different name arrives). For this reason, we have to make the following considerations: It has to be checked for which part (MU) the machine is equipped; if the machine is set up for the same part, then no action is required; if the machine is set up for another part, the set-up time is set anew and the machine starts setting-up after moving the part to the machine.

Attributes and methods for setting up a machine

Method	*Description*
`<path>.setUp`	determines whether a object is currently setting-up(true if the object sets up).
`<path>setUpFor(<mu>)`	triggers setting an object up for a certain MU class. The time needed depends on the value of the setup time for the object.
`<path>.isSetUpFor`	returns the name of the MU (string), for which the object is set up. If the object is not set up for a specific MU, the method returns an empty string.
`<path>.setupTime`	sets/gets the set up time of the object

`<path>.automaticSetUp`	sets/gets whether the set up process is triggered automatically when another MU class arrives at the object

Program the method set up (as the exit control front for the PlaceBuffer):

```
is
   former : string;
do
   -- read the set-up time from the table
   -- set the attribute setupTime
   former:=singleProc.isSetUpFor;
   if former="" then
   --the first part only set-up
   singleProc.setUpTime:=
   setup_times["setup_time",@.name];
   else
   -- former part dismantling_time
   -- recent part setup_time
   singleProc.setUpTime:=setup_times["setup_time",
   @.name] + setup_times["dismantling_time",former];
   end;
   @.move;
end;
```

8.4.2   Methods and Attributes of the TableFile

Delete
Syntax:

```
<path>.delete or <path>.delete(<range>)
```

This method deletes all entries or the specified range in the table/list. The following notation applies to ranges in TableFiles:

- One cell: table[column,row]
- Range: {column1, row1}...{column2, row2}

The range specification consists of two direct specifications, which are separated by two periods. The first specification defines the upper left corner, while the second determines the lower right corner of the range. All entries in the rectangular area will be evaluated. When you enter {*,*} as a second indication, the table evaluates to the largest valid column and row index.

Samples:

{2,2}...{3,3}
{"front","door"}...{"rear","door"}
{3,1}...{*,*}

Notation	Range
{1,2}...{3,5}	*from column 1 to column 3 and row 2 to row 5*
{1,*}...{4,*}	*all rows in column 1 to column 4*
{2,3}...{*,3}	*in row 3 all columns starting with column 2*
{2,3}...{*,*}	*all the columns from column 2 and all rows from row 3*
{*,*}...{3,5}	*all columns to column 3, and all lines to line 5*

Additional Methods of the TableFile are

Method	*Description*
`<path>.copy` `<path>.copy(<range>)`	Copies the contents of the cells to the clipboard
`<path>.initialize` `(<ranges>,<values>)`	Preallocates the specified areas with the passed values, existing contents will be overwritten

`<path>[<column>,<row>]`
`<path>[<column>,`
`<row>]:= <value>;`
	Read/write access; the TableFile allows random access via column and row indices. The index starts and ends with a square bracket. Within the brackets, you first enter the column and then the row of the cell you want to access. If you assign a value, the data type of the value must match the data type of the cell.

`<path>.yDim`	Returns the number of entries (lines).
`<path>.xDim`	Returns the number of columns (which contain values)
`<path>.dim`	returns the product of columns and rows
`<path>.find(<range>,` `<value>)`	Sets the cursor into the cell, which contains the value. You can determine the coordinates with the cursor (table.cursorX and table.cursorY). Before searching make sure that the cursor is located in the correct position. For this you can use `<path>.setCursor(<column>, <row>)`.

| `<path>.insertRow(<value>)` | Adds a new empty row at the position <value> |
| `<path>.writeRow(<position>, <value1>,<value2> …)` | Replaces all entries in the row at the specified position by the passed arguments |

In addition, the TableFile provides methods for inserting and removing columns and rows. You find more information about this in the help and under the heading statistics.

8.4.3 Calculating within Tables

Example 90: Calculating Machine-Hour Rates

Basics: Calculating the machine-hour rate allows a more accurate allocation of common costs and thus a more accurate calculation. The goal is the apportionment of the machine-related indirect production costs to one hour of machine running time. Calculating the machine-hour rate consists of the following components (sample):

Machine-dependent indirect production costs	Fixed amount per month	Variable costs per hour
1. Imputed depreciation[3]	4.500	
2. Imputed interest[4]	900	
3. Imputed rent[5]	500	
4. Energy costs per hour		0.25
5. Tooling costs		10.00
6. Repair/Maintenance		10.00
7. Fuel costs		2.50
Indirect production cost per hour	5900/number of hours machine running time	22.75
Machine-hour rate	5900/number of hours per month + 22.75	

---

[3] Replacement value/asset depreciation range in months.
[4] Cost value/2 * imputed rate/100.
[5] Footprint of the machine in m² * imputed rent per month.

We calculate the machine-hour rate in a TableFile. First, create a sample of the machine-hour rate calculation in the class library (the calculation scheme is the same for all machines, only the values will change). Create the following table:

	string 0	real 1	
string			
1	Replacement value	2000000.00	
2	asset depreciation range in months	120.00	
3	Imputed depreciation		

Calculations in Tables

You can enter values directly into the table cells. Alternatively, you can specify formulas, which calculate the values in the table cells. Tables and lists therefore have two modes: In formula mode, you can enter formulas; in input mode, you can enter values and the values of the formulas are displayed. You can switch to the formula mode with the formula button (to the left of Open):

Calculated fields are shown with a light blue background. A formula has the following basic structure:

PathTableFile[c,r] operator PathTableFile[c,r]

This is a bit cumbersome in relation to the same table, so in calculations within a table you can use the anonymous identifier „?" as a substitute for the path.

Sample:
The imputed depreciation cost is calculated as follows:

Replacement value/asset depreciation range in months. As formula in the table in the example above you would enter:

?[1,1]/?[1,2]

If your formula is wrong, Plant Simulation shows an error message.

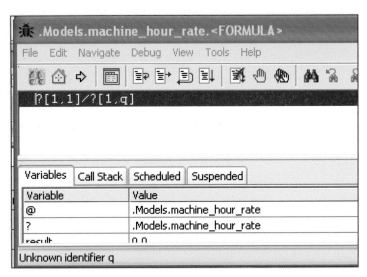

It is important to consider the data types in calculations. The result cell has a certain data type (determined by the data type of the column in the table). The result must also have this data type; otherwise, Plant Simulation will show an error message. A reasonable simplification is the calculation in a single data type (e.g., real) and formatting of the output (e.g., money). The calculation of the machine-hour rate results in the following table (and the associated formulas):

	0	1
1	Replacement value	2000000.00
2	Asset depreciation range in months	120.00
3	Imputed depreciation	?[1,1]/?[1,2]
4	Cost value	1500000.00
5	Imputed rate (%)	7.00
6	Imputed rent	?[1,4]*?[1,5]/200
7	Footprint of the machine in m²	20.00
8	Imputed rent per month and m²	13.00
9	Imputed rent per month	?[1,7]*?[1,8]
10	Total fixed costs per month	?[1,3]+?[1,6]+?[1,9]
11		
12	Energy costs per hour	0.25
13	Tooling costs per hour	10.00

14	Repair/Maintenance per hour	10.00
15	Fuel costs per hour	2.50
16	Total indirect production costs per hour	?[1,12]+?[1,13]+?[1,14]+?[1,15]
17		
18	Monthly machine running time in hours	1.00
19		
20	Machine-hour rate	?[1,10]/?[1,18]+?[1,16]

The cell [1,18] must be calculated in the simulation. The result of the simulation is the actual occupancy of the machine and with the calculation in the table the machine-hour rate (taking account of breaks, occupancy, maintenance, and whatever else you take into account in the simulation).

Create the following Frame:

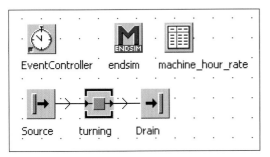

At the end of the simulation, the method EndSim reads (set a month as the end of the simulation in the event controller) the working time of the object turning from the statistics data and writes it to the table machine-hour rate (cell [1,18]). Then, the table calculates the machine-hour rate.

Program the method endSim:

```
is
do
    -- set the machine working time in the table
    machine_hour_rate[1,18]:=
    turning.statWorkingTime/3600;
end;
```

Note:
The values in the table retain their values between the simulation runs. At the beginning of the simulation, you therefore need to initialize all required values (in the example above, the cell [1,8]).

8.5   The TimeSequence

8.5.1   Basic Behavior

You can use the TimeSequence for recording and managing temporary value progressions (stocks, machine output…). The TimeSequence has two columns: Point in time (1st column) and value (2nd column). You can enter values into the Time-Sequence with SimTalk, or the TimeSequence can record values by itself.

8.5.2   Settings

Tab Content

The tab Contents shows the recorded values. You can sort the values in ascending order according to time. The button Set sets empty fields to a default value.

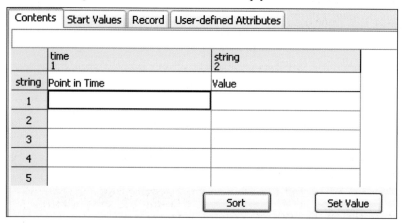

On this tab, you must specify the data types to be stored. This is analogous to the tables:

1. Turn off inheritance (**FORMAT – INHERIT FORMAT**).
2. Then click the right mouse button on the column header of the second column, select Format from the menu.
3. Select the data type, and click OK.

Tab Start Values

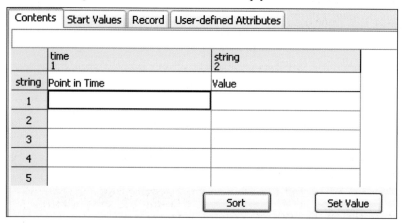

Time reference: You can specify whether Plant Simulation shows time-related data in absolute format (datetime) or in relative format (time).

Reference time: Enter the start of the recording of the values (time, date). The time values are shown relative to this reference value (which shifts the values of the time axis).

Tab Record

Here, you can select the settings that are required for collecting the data.

Value: Enter the relative or absolute path to the value (method, variable, attribute), whose course over time the TimeSequence will record. You might, for example, record the number of parts in the object buffer. The method is buffer.numMU. You can select the value using the button next to the input field methods, attributes, and variables.

Mode: Watch means that values are entered after each change in value. This may possibly lead to a slowdown of your simulation. Sample means that values at certain time intervals are entered (e.g., every 30 minutes). In watch mode, only observable values will be recorded.

Active: Use this to activate or deactivate the TimeSequence.

Example 91: TimeSequence

A process is to be balanced. Three machines supply a fourth machine with parts. The machines M1, M2, M3 have very low availabilities (time-consuming tool testing and adjustments). We are looking for the maximum output of the line, the processing time of M4 and the required buffer size.

Create the following Frame:

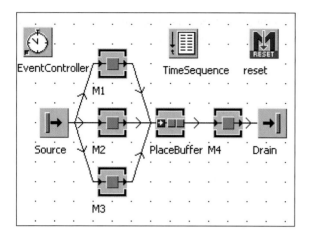

Settings: Source interval 50 seconds, blocking; M1,M2,M3 processing time 1 minute, 50% availability, 45 minutes MTTR; PlaceBuffer capacity 10,000 parts, 0 second processing time; M4 40 seconds processing time, 75% availability, 25 minutes MTTR.

The course of stock in the PlaceBuffer is to be recorded in the TimeSequence.

Follow these steps:

1. Turn off inheritance: Format – Inherit Format (remove the check mark).

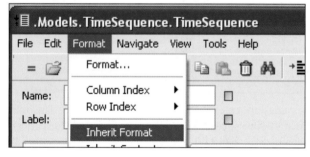

2. Click the tab Record, and select the following settings:

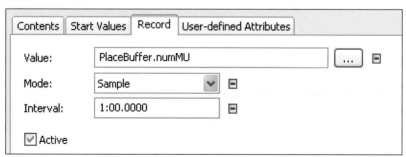

3. Start the simulation. The current time of the EventController and the stock in the PlaceBuffer will be entered into the TimeSequence every minute.

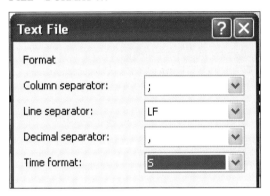

Contents	Start Values	Record	User-defined Attributes	
0.0000				
	time 1		integer 2	
string	Point in Time		Value	
131	2:10:00.0000		20	
132	2:11:00.0000		22	

You can easily export the values of the TimeSequence (e.g., as a text file). First select the format of the text file:

FILE – FORMAT …

Text File	? X
Format	
Column separator:	;
Line separator:	LF
Decimal separator:	,
Time format:	s

Save the table with: FILE – SAVE AS TEXT …

Note:
The EventController does not reset the TimeSequence. You must delete the previous content of the TimeSequence inside a reset or init method.

Example of a reset method:

```
is
do
    timeSequence.delete;
end;
```

The methods and attributes of TimeSequence are those of the TableFile.

8.6   The Trigger

8.6.1   Basic Behavior

The trigger can change values of attributes and global variables during the simulation according to a defined pattern and perform method calls. In addition, the trigger can control a source, so that this starts to produce MUs from a certain moment in time on.

Example 92: Trigger

A machining center produces parts in three shifts (24 h) with a processing time of 1 minute. The following assembly produces one shift with three parallel places, and two shifts with one place. The assembly time is 1:40 minutes. The parts not yet assembled are collected in a buffer. Create the following Frame:

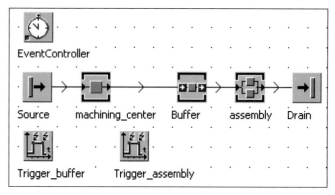

Settings: Source 1 minute interval, blocking; machining_center 1 minute processing time; buffer 0 second processing time, capacity 1000 parts.

After 8 h simulation time, the property assembly.XDim: = 1 (after another 16 hours according to 3 again) must be set. Select these settings in the object Trigger_assembly.

Tab Period

Period	Values	Actions	Representation	User-defined Attributes

☑ Active

Time reference:	Relative	☐
Start time:	0	☐
Active interval:	8:00:00	☐
	☑ Repeat periodically	☐
Period length:	1:00:00:00	☐

Active: Select whether the trigger is active or not during the simulation run.

Time reference: You can select a relative start time (0:00) or an absolute time (date).

Start time: When should be the trigger for the first time active?

Active interval: After what period should be set the value back to the default value (defined in a time line, e.g., 8 hours)?

Repeat periodically: The trigger is active again after the expiration of the period length.

Period length: Sets the duration of a trigger period (e.g., one day or 1:00:00: 00.0000).

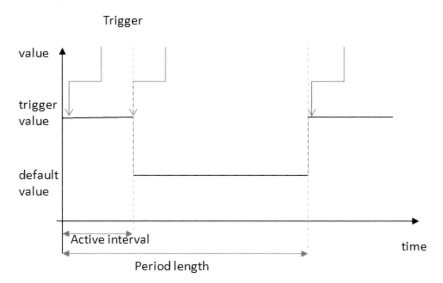

Tab Actions

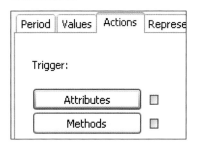

You can start methods or attributes.

Button Attributes: Type into the list which attributes you want to control. An error message appears in the console when Plant Simulation could not execute the action. Before you can type values into the table, you must turn off inheritance and click Apply.

.Models.Trigger.Assembly		
Object	Attribute	Error Message
1 .Models.Trigger.Assembly	xDim	XDim attribute could not be set!

Tab Values

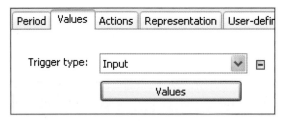

Enter the progress of the value, which the trigger controls, into a TimeSequence.

Button Values:

Before you can type in values, you first have to turn off inheritance in the value table (as table).

Enter the following values in the table:

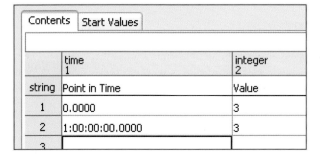

Set the default value (here 1) on the tab Start values.

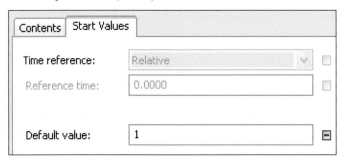

You can check the distribution you set on the tab Representation:

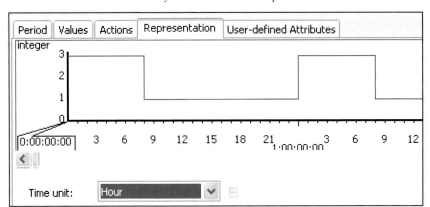

If you run the simulation for a while, you get an error message. Plant simulation cannot reduce the dimension of the parallel station, when parts are located on the respective places. Prior to the reduction of the capacity, the object assembly needs to be emptied. This can, for example, be achieved by temporarily locking the exit of the buffer (e.g., 2 minutes before shift change). The following settings are needed in the object Trigger_buffer: Actions: Attribute buffer.exitLocked, start time: 7:58:00, active interval 2:00, period length one day, repeat periodically, data type Boolean, values 7:58:00 true, default value false.

8.7 The ShiftCalendar

You can use triggers to set the attribute Pause at certain intervals to true or false to model a shift system. It is easier to accomplish this with the object ShiftCalendar. Every material flow object, which "deals with" entities has the following times:

- Planned (working within the shifts)
- Unplanned (times outside the shifts, e.g., weekend)
- Paused (pause within the shifts)

The ShiftCalendar sets these times using a TimeSequence. You can use one Shift-Calendar for the entire simulation, or, in extreme cases, create its own ShiftCalendar for each machine.

Example 93: ShiftCalendar

You are to simulate a continuous process (coating), which has a workplace to prepare and a workplace for follow-up jobs. A coating process takes 8 hours (the facility is 75 m long), the preparing and follow-up job each take 2:30 min. The coating facility works 24 hours a day, 7 days a week. The preparing and follow-up workplaces work according to the following shift system: Beginning of the first shift,

Monday 6.00 clock, end of the last shift: Saturday 6:00 clock. Morning shift start at 6 clock, 14 clock end, break 9:00 to 9:15 clock, 12:00 to 12:30. Middle shift: start 14 clock, 22 clock end, Break 17.00 to 17.15 and 20:00 to 20:30, Night shift: Start 22.00 until 6 clock; breaks analogous to the middle shift. What is the maximum output? Create the following Frame:

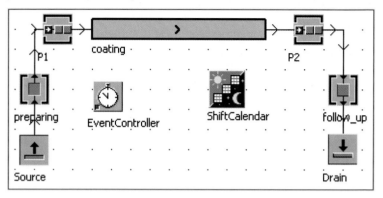

Settings: Length of the entities: 0.2 meters, P1, P2 processing time 0 seconds, capacity 10,000 parts each.

Insert a ShiftCalendar object into the frame. First switch off inheritance on the tab shift times (click on the green icon on the right side + Apply). Then enter the shift times into the table.

	Shift	From	To	Mo	Tu	We	Th	Fr	S	S	Pauses
1	Shift-1	6:00	14:00	✔	✔	✔	✔	✔	☐	☐	9:00-9:15; 12:00-12:30
2	Shift-2	14:00	22:00	✔	✔	✔	✔	✔	☐	☐	17:00-17:15; 20:00-20:30
3	Shift-3	22:00	06:00	✔	✔	✔	✔	✔	☐	☐	01:00-01:15; 04:00-04:30

Tabs: Shift Times | Calendar | Resources | User-defined Attributes

Assign the ShiftCalendar to the objects on the tab Controls – Shift calendar.

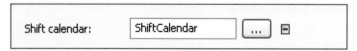

The tab Resources of the ShiftCalendar provides an overview over the stations, which use the ShiftCalendar.

	Objects
1	preparing
2	follow_up

Tabs: Shift Times | Calendar | Resources

8.8   The Generator

The generator starts a method at regular intervals or after a certain time has passed. You can specify all times as a fixed time or as a statistical distribution.

Example 94: Generator, Outward Stock Movement

In the following frame, the produced parts will be not removed by a drain, they will be placed into a store (capacity 10,000). So that the store does not overflow after a short time, we need to simulate outward stock movement. The store will have an average outward stock movement of 80 units per hour. For this purpose, you need a method, which removes 80 parts per hour from the store.

Create the following Frame:

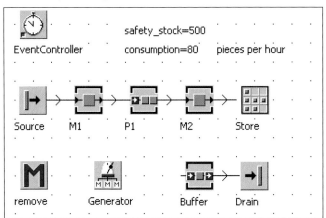

Settings: Source interval 1 minute blocking, M1 and M2 processing time 50 seconds, availability 95%, MTTR 5 hours, P1 capacity 1000 parts, store capacity 10,000 parts. Create safety_stock and outward stock movement (consumption) as global variables, of data type integer, in the frame

Program the method remove (called once per hour):

```
is
   i:integer;
do
   if store.numMU >= (consumption+safety_stock)
      then
      --remove MUs
      for i:=1 to consumption loop
        store.cont.move(buffer);
      next;
   end;
end;
```

In the example above, the method must be called every hour. With the generator you must determine the time and the method, which should be called.

Tab Times

Times	Controls	User-defined Attributes

☑ Active   ☐

DDD:HH:MM:SS.XXXX

Start:	Const ▾	1:00:00	⊟
Stop:	Const ▾	0	☐
Interval:	Const ▾	1:00:00	⊟
Duration:	Const ▾	0	☐

Active: Activate the generator.

Start: Select when the interval control will be activated for the first time.

Stop: Select at which simulation time no interval control should be active.

Interval: What time should elapse between calls?

Tab Controls
Select your method on this tab:

Times	Controls	User-defined Attributes

| Interval: | remove | [...] ⊟ |

8.9   The AttributeExplorer

You can manage a variety of attributes of different objects from a single central location with the AttributeExplorer.

Example 95: AttributeExplorer

Create the following Frame:

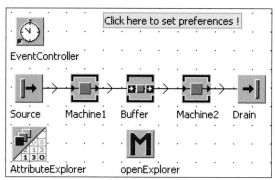

Insert a comment with the text "Click here to set preferences!" (to open the At-tributeExplorer). The processing times of Machine1 and Machine2 should be changed in a single dialog box. Therefore, add an AttributeExplorer to the frame. Open the AttributeExplorer by double-clicking it. Select the tab Objects and turn off inheritance (+ Apply). Drag the items from the frame to the list of object paths. Press Enter after each object to add a new line to the table.

Data	Objects	Attributes	Query	User-defined Attributes
				Objects
1			.Models.AttributeExplorer.Source	
2			.Models.AttributeExplorer.Machine1	
3			.Models.AttributeExplorer.Buffer	
4			.Models.AttributeExplorer.Machine2	

*Next, select the attributes that you want to view and modify. In the example above, these are the attributes interval (source), procTime (Machine1 and Machine2), and capacity (buffer). Enter these settings on the tab **ATTRIBUTES**. Use the column alias to display a different name than the attribute name in the AttributeExplorer (for instance processing time instead procTime). You can select the attribute with the button Show attributes.*

   First, turn off inheritance and click Apply.

Data	Objects	Attributes	Query	User-defined Attributes
	Name			Alias
1	Interval			
2	ProcTime			Processing time
3	capacity			

If you want to display the alias names, select the option Show attributes with alias on the tab Data. You can also select to show the paths or the labels of the objects.

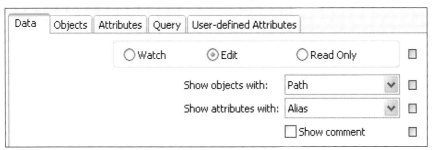

Click the button Show Explorer to open a window in which you can set all values at once

	Interval	Processing tir	capacity
.Models.AttributeExplorer.Source	2:00.0000	0.0000	1
.Models.AttributeExplorer.Machine1		1:00.0000	1
.Models.AttributeExplorer.Buffer		0.0000	4
.Models.AttributeExplorer.Machine2		1:00.0000	1

The AttributeExplorer itself should open when you click the comment ("Click here ..."). Open the comment. Select Tools – Select controls.

.Models.AttributeExplorer.Comment

Navigate   View   Tools   Help

Select Controls...

Name: Comm

Select the following:

Select: openExplorer

Program the method openExplorer:

```
is
do
   -- activate the AttributExplorer
   attributeExplorer.Active:=true;
end;
```

When you now click on the comment, the AttributeExplorer opens. An issue results in this way in older versions of Plant Simulation. You then cannot open the dialog of the comment by double-clicking (before release 9). You can open the dialog via the structure of the frame. Click the right mouse button on the frame in the class library. Select Show structure from the menu. You can then double-click the objects in the opening window and thus open their dialogs.

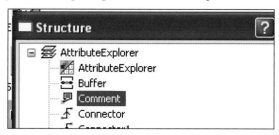

Structure   ?

AttributeExplorer
   AttributeExplorer
   Buffer
   Comment
   Connector

8.10 The EventController

The EventController enables access to the system time. Furthermore, you can call all buttons of the EventController in SimTalk.

Methods of the EventController:

Method	Description
`<path>.SimTime`	Returns the current simulation time (data type time).
`<path>.AbsSimTime`	Returns the current simulation time (data type datetime).
`<path>.start`	Starts the simulation
`<path>.step`	
`<path>.stop`	
`<path>.reset`	

9 Statistics

9.1 Basics

Most questions in simulations deal with distributions of, for example, failures, waiting times, etc. For the evaluation of simulation runs, you can rely on statistical data, which the material flow objects collect.

9.1.1 Statistics Collection Period

The statistics collection period is the time interval between activating the collection of statistical data and the query of statistics. Statistics data are recorded only if the collection of statistics in the objects is active. If statistics is disabled, all statistical data of an object will be deleted. You can reset statistics collection in the EventController at a certain time. In this way, you can hide the ramp-up behavior of your model, and statistics collection can start when the system reaches full output. You can enter this setting in the EventController, on the tab Settings. Enter the time at which the event controller will reset statistics into the field STATISTICS.

Controls	Settings	
Date:	2009/01/06 00:00:00	
End:	11:00:00:00	
Statistics:	1:00:00:00	

With the setting above, the EventController will reset statistics of the objects after one day. When the simulation is finished after 11 days, the objects recorded statistical data for 10 days. The following description illustrates the composition of the statistics collection period (scheduled time only):

Statistics collection period					
Resource not paused					R. paused
Resource operational			Resource not operational		
Waiting time	Set-up time	Working time	Blocked time	Failed time	

Working time: A resource works when at least one MU is being processed on the object (setup times and recovery times are not included in the working time).

S. Bangsow: Manufacturing Simulation with Plant Simulation, Simtalk, pp. 223–252, 2010.
© Springer Berlin Heidelberg 2010

Failed time: A resource is failed, if it is not paused and its attribute failed has the value True.

Blocking time: A resource is blocked if it

- is full
- neither failed nor paused and
- all places do not work (e.g., the MU is processed and cannot be passed on).

Paused: A resource is paused when its paused attribute has the value True. In addition, more than 100 different values for statistical analysis are available.

9.1.2   Activating Statistics Collection

Open a material flow object (double-click) – click the tab Statistics. Enable Resource statistics, and click OK or Apply.

You can also enable or disable statistics collection of an object using the method

```
<path>.ResStatOn:=true;  --or false
```

Note:
By default, statistics collection is turned on for all material objects. To increase the performance of the simulation, it can help to deactivate statistics collection for all objects for which you do not need statistical analysis.

9.2   Statistics – Methods and Attributes

You can read all statistics data with SimTalk.

Method	Description
`<path>.statistics`	Shows statistics of the object on screen
`<path>.statistics(<table>)`	Statistics will be written in the specified Plant Simulation table.
`<path>.statistics(<string>)`	Statistics will be written in the specified file.
`<path>.statWaitingPortion`	Returns the percentage of the waiting time relative to the total time (data type real).

`<path>.statWorkingPortion`
`<path>.statBlockingPortion`
`<path>.statFailPortion`
`<path>.statSetupPortion`
`<path>.statPausingPortion`
`<path>.statUnplannedPortion`
`<path>.statEmptyPortion`
	see above

`<path>.statNumIn` `<path>.statNumOut`	Returns the number of MUs that entered (leaved) the object. The returned data type is integer.
`<path>.statMaxNumMU`	Returns the maximum number of places occupied during the simulation.
`<path>.initStat`	The method initStat resets the statistics of an object. This can be useful if the statistics recording should not start at the beginning of the simulation, the statistics collection will contain only values from the call of initStat.

Example 96: Statistics

You are to simulate a manufacturing cell. Four machines feed a chemical treatment unit. The chemical treatment unit has a constant feed rate of 0.00333 m/s. The part has a length of 0.4 m. The machines have a processing time of 4 minutes, an availability of 50% and a MTTR of 3 hours. A buffer with capacity of 100 parts is located in front of the chemical treatment unit, which is 30 meters long. The source produces parts at an interval of 2 minutes. Create the following Frame:

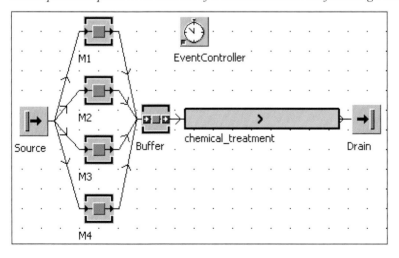

This example will show some typical statistical analysis.

1. The statistical data is to be written into a file at the end of the simulation. This can be easily accomplished with a table. During or at the end of a simulation, you write the statistics data into a table and then save the table as a file. For this the TableFile provides the following methods:

Method	Description
`<path>.writeFile(<string>)`	This method writes the contents of the table into a text file. Pass the path as argument. Existing files with the same name will be overwritten.
`<path>.writeExcelFile(` `<string>,[<string>])`	This method writes the contents of the table into an Excel file. Pass the filename (path). As a second argument you can pass an Excel table name. To be able to use this method, MS Excel has to be installed on your computer

To simplify statistical analysis, write a method, which writes the statistical data of all material flow objects to a TableFile.

Add a TableFile "analysis" to the frame. Format the table according to the following example:

	string 0	real 1	real 2	real 3	real 4	real 5	
string		working	waiting	blocked	failed	paused	
1							
2							
3							

Preliminaries: With the help of the frame object, you can access all objects in the frame by an index (method: <path>.node(<integer>)). The method <path>.num-Nodes returns the number of objects within the frame. Finally, you can query the class of the objects with <path>.class (e.g., MateriaFlow.SingleProc).

Add an endSim-method to the frame. The method iterates through all objects in the frame. For objects of class SingleProc, the method inserts a row into the table and writes its name and statistical values into it. Finally, the method exports the contents of the TableFile to an Excel file: simulation_analysis.xls.

Method endSim:

```
-- writes statistical data for all SingleProc
-- objects
is
   i:integer;
   obj:object;
```

```
do
  analysis.delete;
  for i:=1 to current.numNodes loop
    obj:=current.node(i);
    if obj.class = .MaterialFlow.SingleProc then
      current.analysis.writeRow(0,
      current.analysis.YDim+1,
      obj.name,obj.statWorkingPortion,
      obj.statWaitingPortion,
      obj.statBlockingPortion,
      obj.statFailPortion,
      obj.statPausingPortion);
    end;
  next;
  -- write  excel file
  current.analysis.writeExcelFile(
  "c:\simulation_analysis.xls");
end;
```

You can easily extend this method by adding other classes. Set a simulation end of 2 days, and let the simulation run up to this point. Search for the Excel file on your "c:\" drive.

	A	B	C	D	E	F
1		working	waiting	blocked	failed	paused
2	M1	0,28	0,02	0,16	0,54	0,00
3	M2	0,33	0,03	0,12	0,52	0,00
4	M3	0,29	0,03	0,14	0,54	0,00
5	M4	0,24	0,02	0,19	0,56	0,00

2. Determining Average Values

Often average values need to be calculated within a simulation. A typical example is the calculation of average stock. For an average calculation, you need a series of values and the number of the values (arithmetic mean). Within the simulation you can choose another approach. A generator calls a method every hour that determines the number of parts within the frame. The method calculates a new average based on the old average, the number of hours, and the new stock. You can easily find out the number of MUs in the frame with <path>.NumMu. (Keep in mind though that the method numMU also counts containers and transporters.)

Add the variable average_stock (real) to the frame. Insert a generator into the frame and a method "new_stock". The generator calls "new_stock" once per hour starting after 1 hour.

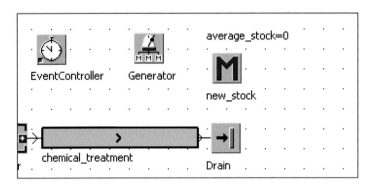

Program the method new_stock:

```
is
   hours:integer;
do
   hours:=time_to_num(eventController.simTime)/3600;
   average_stock:=
   (average_stock*(hours-1)+ current.numMu)/hours;
end;
```

3. Record Values

For the evaluation and optimization of the simulation, you often need the progression of values over time. You can accomplish this in two ways:

a) Record values with TimeSequence objects (a separate TimeSequence for each value).

b) Record values with a table and analysis of the table.

Example: In the example above, you are to show the distribution of the failures of the individual machines (hourly). If a failure has occurred, the value 1 should be entered, if the object is not failed at the moment, the value 0 should be entered. Insert the table failure_machines into the frame and format it as follows:

	time 1	integer 2	integer 3	integer 4	integer 5
string	time	M1_failure	M2_failure	M3_failure	M4_failure
1					
2					
3					

Program the method record_failures:

```
is
  i:integer; -- next entry
do
  i:=failure_machines.YDim+1;
  -- call once per hour
  failure_machines[1,i]:=eventController.simTime;
  -- set values depending of attribute failed
  if M1.failed then
    failure_machines["M1_failure",i]:=1;
  else
    failure_machines["M1_failure",i]:=0;
  end;
  if M2.failed then
    failure_machines["M2_failure",i]:=1;
  else
    failure_machines["M2_failure",i]:=0;
  end;
  -- and so on
end;
```

4. Data Collected by the Drain

The drain collects detailed statistics about the destroyed parts. Open the drain, and select the tab: Type Statistics.

Times	Set-Up	Failures	Controls	Statistics	Type Statistics	User-defined Attributes

☑ Type dependent statistics

Detailed Statistics Table

Working:	96.43%	Average lifespan:	9:10:54.3113
Delayed:	2.11%	Average exit interval:	4:00.2402
Setup:	0.00%	Total throughput:	7155
Failed:	1.46%	Throughput per hour:	14.99
Paused:	0.00%	Throughput per day:	359.61

Click the button Detailed Statistics Table to receive further information.

Type Statistics

	Type	Time	Total through	%Parts	LT_Mean	LT_StdDev	LT_Min
1	Entity	20:00:00:00.I	7155	100.00	9:10:54.3113	1:11:29.6262	2:34:09.009

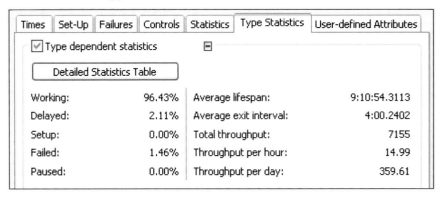

The drain provides a number of methods for accessing statistics; here is a small selection:

Method	Description
`<path>.typeStatistics(<table>)`	Copies the type statistics table in the specified table
`<path>.typeStatisticsCumulated(<table>)`	Analogous
`<path>.statThroughputPerHour`	Returns the throughput per hour (real)
`<path>.ThroughputPerDay`	Returns the throughput per day
`<path>.statAvgLifeSpan`	Returns the average throughput time

Most statistical data can be understood more easily if they are presented graphically. For this purpose, Plant Simulation provides user interface objects.

9.3   User Interface Objects

9.3.1   Chart

The object Chart represents data in Plant Simulation graphically. In watch mode, the graphic is automatically updated after each modification of a displayed value. (The value must be observable.) This way you can visualize the dynamic behavior of certain values during the simulation.

9.3.1.1 Plotter

In the following example, the development of stock in a buffer is to be presented graphically.

Example 97: Plotter

Create the following Frame:

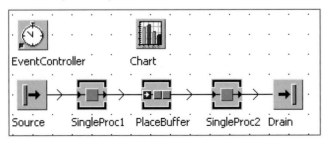

Settings: SingleProc1, SingleProc2 processing time 1 minute each, availability SingleProc1 and SingleProc2 95% MTTR 3 hours, use different random streams, PlaceBuffer capacity 100, accumulating, 30 seconds processing time.

There are two ways to present data in charts: From input channels or from Table-Files.

Input Channel
The Chart object itself records and displays the data. You can access the recorded values using SimTalk.

Example: You are to display the stock in the PlaceBuffer.

1. Click the tab Data in the Chart and select the **DATA SOURCE – INPUT CHANNELS.**

Clicking the button opens a table into which you can enter the name of the object, the path of the displayed value and comments. First, turn off inheritance of the table (+ Apply). Then click the button **TABLE FILE.** *Enter the path and the attribute, which is to be displayed (PlaceBuffer.numMU in our example.)*

	string 0	string 1		string 2
string		stock		
1		PlaceBuffer.numMU		
2				

You still need to select at which interval Plant Simulation is to update the chart on the tab Data. Watch mode updates after every change of the observable value, sample mode updates within the set interval, and plot mode updates the graph after each simulation event.

Example: The chart will be updated every minute, the setting for this is as follows:

Mode:	Sample	☑
Interval:	1:00	☐

2. Select the category plotter and Chart type line on the tab Display.

Data	Display	Axes	Labels	Color	Font	User-defined A

Category:	Plotter	∨ ⊟
Chart type:	Line	∨ ☐
3D effect:	(None)	∨ ☐

If you select the option Display in frame, Plant Simulation shows the diagram (with its values) instead of the object's icon in the frame.

*3. You then have to select some settings on the tab **AXES**. Enter the number of displayed values (e.g., 10,000). The scroll bar option is mostly useful for presentations using a plotter. In addition, you have to enter the size of the displayed time range in the plotter window in the box **RANGE X**, enter 1:00:00:00 (one day).*

Number of values:	10000	☑ Scrollbar	⊟	
Range: Y:	0	...	*	☐
X:	1:00:00:00	Feed rate (grid units):	6	⊟

*4. Add labels to the plotter on the tab **LABELS**. Here you can also specify whether a legend is displayed and how it is displayed in the chart window.*

Data	Display	Axes	Labels	Color	Font	User-defined Attributes

Title:	Buffer stock	⊟
Subtitle:		☐
X-axis:	Time	⊟
Y-axis:	MUs	⊟
Legend:	(Off) ∨	☐ Annotations ☐

*You can select additional format settings on the tabs **FONT AND COLOR**. Clicking the button **SHOW CHART** opens the window of the plotter:*

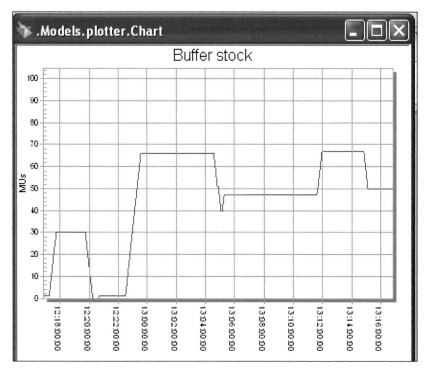

You can access the data, which the plotter records, and write them into a TableFile.

Example: Add a TableFile (analysis) and a method (saveData) to the frame. You can read the data of the chart object with the method:

```
<path>.putValuesIntoTable(<table>)
```

First, turn off inheritance for the TableFile.

Program the method saveData:

```
is
do
  -- delete previous values
  analysis.delete;
  -- write the chart data into the table analysis
  chart.putValuesIntoTable(analysis);
end;
```

Plant Simulation formats the table and inserts the data of the plotter.

9.3.1.2 Chart Types

Plant Simulation provides a large number of different chart types to display values.

Example 98: Chart from a TableFile

Continuing the example above:

You are to show the composition of the statistics collection period of the objects SingleProc1 and SingleProc2. You want to write the values into a TableFile. You want to show the values of the table in a chart. The following values will be displayed:

Value	SimTalk Attribute
Waiting time	`<path>.statWaitingPortion`
Working time	`<path>.statWorkingPortion`
Blocked time	`<path>.statBlockingPortion`
Failed time	`<path>.statFailPortion`
Paused time	`<path>.statPausingPortion`

Insert a TableFile with the following formatting (chart_values) into the Frame:

	string 0	real 1	real 2	real 3	real 4	real 5
string		waiting	working	blocked	failed	paused
1	SingleProc1					
2	SingleProc2					

At the end of the simulation run, a method (endSim) will write the statistics values of the two SingleProc into the TableFile chart_values.
Program the method endSim:

```
is
do
   -- values of SingleProc1
   chart_values.writeRow(1,1,
     SingleProc1.statWaitingPortion,
     SingleProc1.statWorkingPortion,
     SingleProc1.statBlockingPortion,
     SingleProc1.statFailPortion,
     SingleProc1.statPausingPortion);
   -- values of SingleProc2, next row
   chart_values.writeRow(1,2,
     SingleProc2.statWaitingPortion,
     SingleProc2.statWorkingPortion,
     SingleProc2.statBlockingPortion,
     SingleProc2.statFailPortion,
     SingleProc2.statPausingPortion);
end;
```

Creating Charts

*1. Insert a Chart object into the frame. Open the chart and select the Data source – TableFile on the tab **DATA**. Enter the name of our table, chart_values, into the field **TABLE**:*

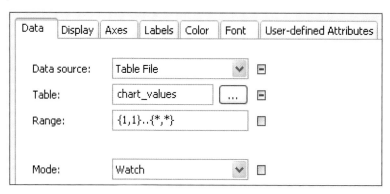

*2. Select **CATEGORY** > **CHART** on the tab **DISPLAY**. The chart-type stacked bars (100%) is suited for displaying the portions of the statistics collection period. You need to set **DATA – IN COLUMN** (structures of the table chart_values – the data are spread across several columns).*

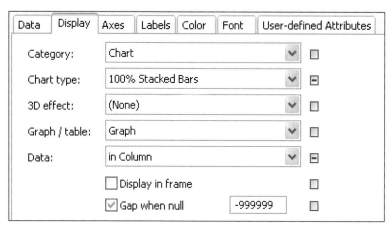

*3. Label your chart. Show the legend on tab **LABELS**. To avoid misunderstandings, you must customize the colors in the chart so that they match the colors of the status LEDs (in any case, failures should be shown red, pauses blue, and blockages orange). You can set the order of colors on the tab **COLOR**. Double-click a color to change it, and then select a new color. The order of colors in the example above must be yellow, green, orange, red, and blue.*

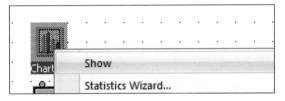

Data	Display	Axes	Labels	Color	Font	User-define

Color	Line Style	Line Weight	Marker Type
1	_____	Thin	Solid Circle
2	_____	Thin	Solid Square
3	_____	Thin	Solid Diamond
4	_____	Thin	Solid Triangle Up

The chart will be shown by clicking the button Show chart. You can also show it by using the context menu of the chart object icon.

Chart type-Stacked bars:

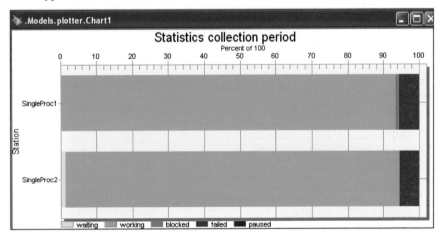

9.3.1.3 Statistics Wizard

If you want to analyze the statistics collection period of all objects of a certain class, you can use the statistics wizard. For that, add a chart object to the frame. Click the right mouse button on the chart icon in the frame. Select **STATISTICS WIZARD** from the context menu.

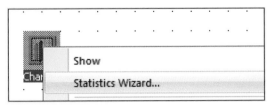

Select the objects whose statistics you want to show in the dialog of the statistics wizard. Leave SingleProc and Production checked.

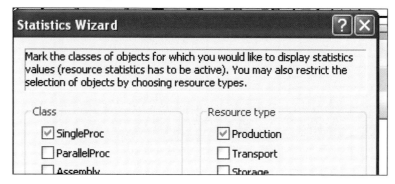

9.3.1.4 Histograms

Histograms show the frequency of certain values in relation to the simulation time.

Example 99: Histogram

We continue with the example above: You are to display the distribution of the occupancy of the PlaceBuffer (attribute numMU).

1. Add a new chart to the frame. Select **DATA** *–* **INPUT CHANNELS** *and enter Place-Buffer.numMU in the table.*

	string 0	string 1	string 2
string		Stock	
1		PlaceBuffer.numMU	

2. Select **CATEGORY** *–* **HISTOGRAM** *and the* **CHART TYPE** *–* **COLUMNS** *on the tab* **DISPLAY**.

Data	Display	Axes	Labels	Color	Font	User-defined Attribute

Category:	Histogram	▼	⊟	☐ Acc
Chart type:	Columns	▼	⊟
3D effect:	(None)	▼	☐
Graph / table:	Graph	▼	☐

*3. Clicking **SHOW CHART** displays the histogram.*

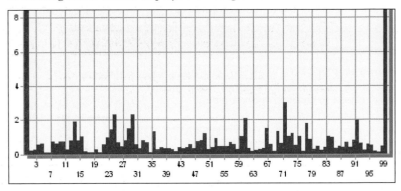

> **Note:**
> You can copy charts in the dialog box of the chart object to the clipboard and then paste them into other programs (e.g., PowerPoint). Select Tools – Copy to Clipboard in the dialog of the Chart.

9.3.2   The Sankey Diagram

The Sankey diagram is used for visualizing the distribution of the material flow. For this, Plant Simulation uses lines with different widths. The Sankey diagram is located in the folder Tools, or on the toolbar Tools.

Example 100: Sankey Diagram

The following frame shows how the Sankey diagram works:

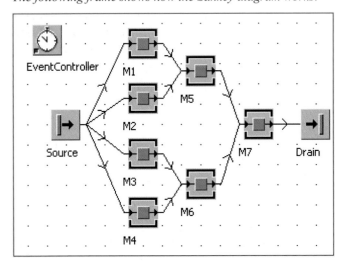

Settings:

Machine	Processing time	Availability	MTTR
M1	1:00.0000	95%	2:00:00.0000
M2	1:00.0000	85%	2:00:00.0000
M3	1:00.0000	70%	2:00:00.0000
M4	1:00.0000	50%	2:00:00.0000
M5	1:00.0000	95%	2:00:00.0000
M6	1:00.0000	85%	2:00:00.0000
M7	50.0000	95%	2:00:00.0000

The source produces parts with an interval of 1 minute (blocking), the exit strategy is cyclic blocking. Add a SankeyDiagram to the frame. Open the SankeyDiagram by double-clicking it. Click the button Open (MUs to be watched).

Enter the MU class, which is to be observed, into the following table. Drag the class Entity from the class library into the table.

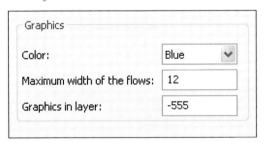

You can select some formatting options, such as color settings and the maximum width of the streams:

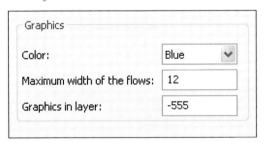

Graphics in layer determines the z-position of the Sankey display. The smaller the number, the closer to the foreground a graphic is located. Finish your settings by clicking OK. Now, run the simulation for a while (50 days). Then click the right mouse button on the object SankeyDiagram in your frame. Select DISPLAY SANKEY DIAGRAM. DELETE SANKEY DIAGRAM deletes the Sankey streams.

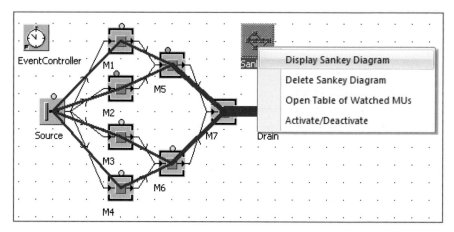

The thicker the Sankey streams between two stations, the more MUs have been transported on the connectors or methods between these stations. The exit strategy cycle of the source leads to the stations M1 to M4 receiving the same number of parts. If a machine fails, the source waits with the transfer process until the machine is operational again. M1 to M4 receive the same number of parts (Sankey lines have the same width). Output after 50 days is 21,230 parts.

You are to simulate a second variant. Click the right mouse button on the Frame in the class library, and select Duplicate. Close the frame window, and open the duplicate. Change this frame as follows:

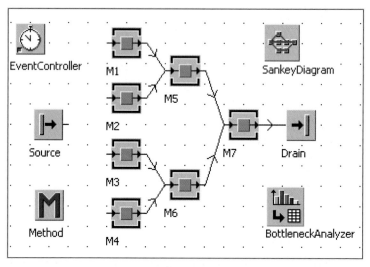

Program the method and assign it as the exit control (front) to the source. The source is to transfer the parts to the first available and operational machine.

Method:

```
is
do
  waituntil (m1.operational and m1.empty) or
  (m2.operational and m2.empty) or
  (m3.operational and m3.empty) or
  (m4.operational and m4.empty)  prio 1;
  if (m1.operational and m1.empty) then
    @.move(m1);
  elseif (m2.operational and m2.empty) then
    @.move(m2);
  elseif (m3.operational and m3.empty) then
    @.move(m3);
  elseif (m4.operational and m4.empty) then
    @.move(m4);
  end;
end;
```

Run this simulation for 50 days. The output is about 72,000 parts. The Sankey diagram now reflects the availability of the machines (the lower the availability, the fewer parts run across the machines).

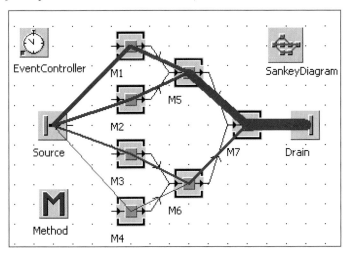

9.3.3   The Bottleneck Analyzer

The Bottleneck analyzer visualizes the default statistics for all selected objects. It is quite simple to use. First, make sure that sufficient space is available above the top object. Open the object BottleneckAnalyzer and click the tab Configure. Select object types for which you want to display statistics.

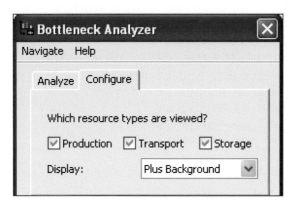

Click the button Analyze on the tab Analyze to create the statistics evaluation.

The statistical data is displayed graphically in the frame. You can also output the data after the analysis as a table. Click **RANKING TABLE – OPEN**. Once you have chosen a sorting option, the table is displayed.

Sorted according to working time				
root.M7				
	object 1	real 2	real 3	real 4
string	resource	working	setup	wait
1	root.M7	83.33	0.00	
2	root.M5	70.81	0.00	
3	root.M1	39.62	0.00	

9.3.4   The Display

9.3.4.1 Behavior

You can use the display object to show dynamic values (attributes, variables) during a simulation run. The values can be represented as a number or bar. The activated object periodically checks the value and updates the display (Sample mode) or after a corresponding change (Watch mode). As a bar or pie, the display shows numeric values in relation to the specified interval (between min and max).

Example 101: Display

Create the following Frame:

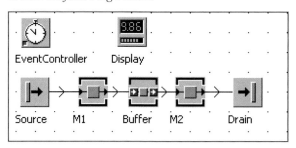

Settings: Source interval: 2:10, M1 processing time: 2:00 availability: 90% 1 hour MTTR, M2 1 minute processing time 50% availability, 2 hours MTTR, Buffer capacity 1,000 no processing time. The display should show the stock of the buffer.

9.3.4.2 Attributes of the Display

Tab Data

Data	Display	User-defined Attributes

Path: `.Models.display.Buffer.numMU`

Comment: `Buffer stock`

Mode: `Sample`

Interval: `1:00`

Path: Enter the path to the observed value (relative or absolute). You can enter global variables, attributes, and methods (invalid paths are marked).

Comment: Enter a detailed description of the Display that is displayed under the object.

Mode: Select Watch or Sample mode (with interval).

Tab Display

The value of the display can be displayed as a bar (numeric values) or pie, or as text.

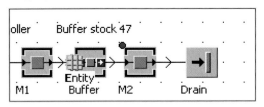

Display as text:

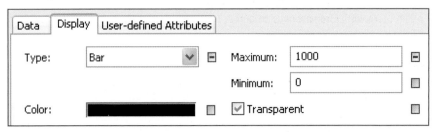

You can only adjust the color and font size and set a transparent background for the display.

Display as a bar/pie

The bar/pie shows the ratio of the actual value and a given maximum value.

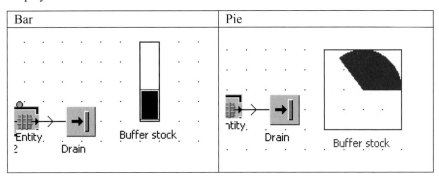

Display:

Bar	Pie

9.3.5   The Comment

The comment has no active behavior during the simulation run and can be used for explanations and labeling.

Tab Display

Text: The text, which you enter here, will be shown in the frame. You can assign the text dynamically using the method <path>.text:= <string>.

Font size, font color, background color: Select formatting options for the text of the comment. If Transparent is selected, the comment is shown within a box with the background color shining through.

Tab Comment

In the big text box on the tab Comment, you can save more text which only is visible after opening the comment object. This text can be created in Rich Text Format (e.g., you create the text in Word and paste it into the comment via the clipboard). You can find formatting options in the context menu of the input field. You can access the contents of the comment with <path>.cont.

9.3.6   The Report

A report can present a very large number of data. The report consists of header data and the report data, which you can arrange hierarchically.

9.3.6.1 Automatic Resource Report (Statistics Report)

You can automatically create reports in Plant Simulation. Select the objects for which you want to create a report by holding down the Shift key. Then press the F6 key.

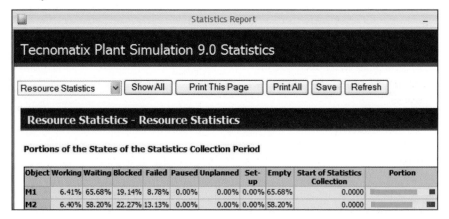

9.3.6.2 Report Header
Example 102: Report

Use the example statistics for creating of the report. Insert a report into the frame, and open the report by double-clicking it. Type in general information about the simulation on the tab General.

Structure	General	User-defined Attributes

Name of simulation run:	Chemical treatment
Version of simulation run:	2.10
Person in charge:	John Smith
Save to folder:	
Save under file name:	
Save as type:	Web Page, complete (.htm;.html)
Window height:	600 ☐ [Pixel] Window width: 640

These data are shown later in the report header.

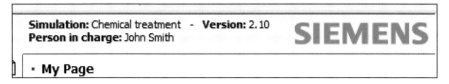

9.3.6.3 Report Data

Define the structure of the report and the displayed information on the tab Structure. In the left pane, you can add using the context menu each (fold) object new pages. You must first turn off inheritance. Click the right mouse button on Report and select **NEW** from the context menu:

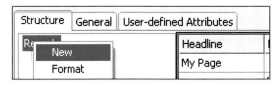

Rename the first sheet of the report to General. Click the entry with the right mouse button. Select **RENAME** from the context menu. Then you can overtype the name of the page. The data of each page will be structured using headings. Each page is separated into three columns. Each column can either contain an icon, text, or an object call.

Example: The first page is to include a brief description of the simulation and a screenshot of the frame. Type "General" in the box Headline, then double-click in the box next to it (column 1). You can place text, icons, or method calls in the report. Select the format text for the first field.

Type the following text into the box Show object: "It was to simulate a chemical treatment, which is supplied by four machines. Through a complex work process, the machines have an availability of only 50%."

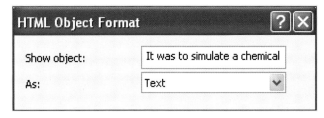

The next line is to remain free. You can insert HTML tags as text; the HTML command for a blank line is <br>. The report consists of HTML pages. Embedded HTML instructions accordingly modify the appearance of the report. Type the following into the second row, first column:

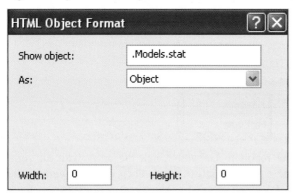

Below is to be shown a screenshot of the frame. Pressing Enter in the last row, last column of the table on the tab structure creates a new row. Enter the following into the first column of the third row:

HTML Object Format

Show object: .Models.stat

As: Object

Width: 0   Height: 0

Enter the address of your frame in the class library. Set width and height both to zero. Plant Simulation then determines the width and height of the image. The structure of the report page should look as follows:

Headline	First column	Second column	Third column
General	It was to simul...		
	\<br\>		
	.Models.stat		

You can view the report by clicking the button Show Report.

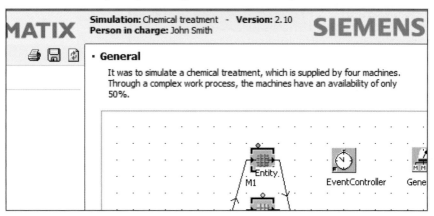

9.3.6.4 Texts in Reports

You can enter text directly into the report (see above) or use the comment object for inputting text and output the contents of the comment block in the report.

Example: Insert a comment object into the frame, and type the following into the tab Comment:

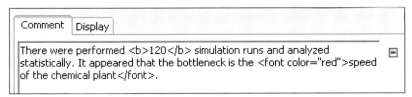

Disable the option in the comment object:

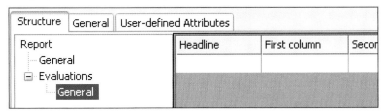

Add a page to the report: evaluations. Then click the right mouse button on the new page and add another page with New (General).

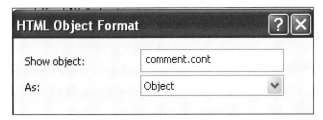

The title of the page should be "Evaluations". Enter the following into the first column of the first row:

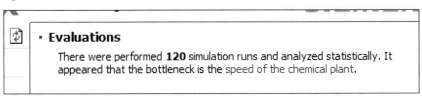

The content of the comment is shown with the included HTML formatting in the report.

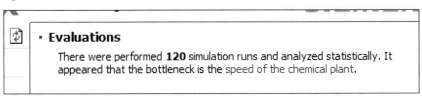

You can use the following HTML commands for formatting your text (selection).

HTML-tag	Description
`<H1> Headline 1 </H1>`	Headline outline levels 1 to 6
`<BIG>Text </BIG>`	Rel. enlarged text
`<SMALL>Text </SMALL>`	Smaller text
`<FONT SIZE=7>Text </FONT>`	Font size (1 very small to 7 very large)
`<FONT COLOR="red">Text </FONT>`	Font color as color or hexadecimal
`<FONT FACE="Arial, Tahoma"> Text</FONT>`	Font face as list separated by commas
`<B>Text</B>`	Bold
`<I>Text</I>`	Italic
`<U>Text</U>`	Underlined text
`<S>Text</S>`	Strike through
`<SUB>Text</SUB>`	Subscript
`<SUP>Text</SUP>`	Superscript
Combination: `<B><I><U>` bold, italic, underlined `</U></I></B>`	

9.3.6.5 Show Objects in Reports

You can access attribute values in reports and display objects. When you display objects in reports, Plant Simulation creates an image of the object (e.g., graph, Frame), or displays values of default attributes. You can type in a complete Sim-Talk call into the field **SHOW OBJECT**.

Example: Insert the page "Statistics" into the report. Here you are to display the main statistics of the machines. The headline is the name of the machine. The first column should display the names of the values in the second column and in the third column the units of the values. At the end of the page, a chart with the statistical data is to be displayed. To display the value of statWorkingPortion in the second column, you need to enter the following settings:

Show object: round(M1.statWorkingPortion*100,2)

Show as Object

Note:
The default setting in the report is Show as Object. If you want to display text, you must switch to Show Object as Text, otherwise you get an error when calling the report.

The settings for displaying statistical data for machine M1 look as follows:

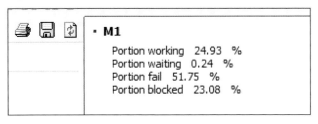

Structure	General	User-defined Attributes			
Report		Hea...	First column	Second column	Thir(
General		M1	Portion working	round(M1.statWorkingPortion*100,2)	%
Evaluations			Portion waiting	round(M1.statWaitingPortion*100,2)	%
General			Portion fail	round(M1.statFailPortion*100,2)	%
Statistics			Portion blocked	round(M1.statBlockingPortion*100,2)	%

Report:

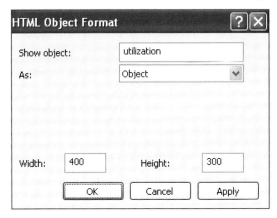

```
· M1
     Portion working   24.93  %
     Portion waiting    0.24  %
     Portion fail   51.75  %
     Portion blocked   23.08  %
```

Charts and tables are inserted into the report via a simple object call. When inserting a chart you must specify a size for displaying it. Example: The chart object has the name "utilization", the necessary setting in the report looks as follows:

HTML Object Format

Show object:	utilization
As:	Object
Width:	400
Height:	300

OK   Cancel   Apply

To display methods, we have to use a little trick. You can access the text of the method as follows:

```
ref(<path>).program
```

The attribute program returns the entire text of the method, including the control characters. The control characters are normally ignored in the HTML display, so there will be a presentation without line breaks and tabs. With the HTML statement <pre>text</pre>, you can force the Report to display line breaks and control characters. If you want to display a method in the report, use the following setting (display as object):

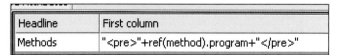

Headline	First column
Methods	"<pre>"+ref(method).program+"</pre>"

9.3.6.6 Show Images in Reports

You can also display images in reports. The images must be created as icons of the report (Context menu – **EDIT ICONS – ICON – NEW – FILE – OPEN** …). You must specify the icon number when inserting it into the report, for example, the image is saved as an icon in the object Report (icon No. 16):

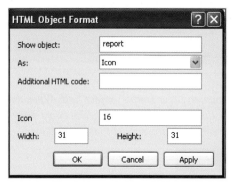

10 User Interface Objects

10.1 General

You can use the object dialog to create your own dialog boxes. You can create dialogs for your own objects for the user of the simulation model to facilitate the operation of your objects. The dialog can be used to select the settings in complex frames and subframes. It serves as an interface between the simulation and the user. In this way, you can create simulations, which can be operated by users who have no knowledge of Plant Simulation.

10.2 Elements of the Dialog

Each dialog object manages a single dialog box. A dialog may consist of the following basic elements:

- Comments/labels
- Text fields
- Buttons, menus
- List-Boxes
- Radio buttons and checkboxes
- Tabs
- Images
- Tables

Example 103: Dialog

The processing time and the failures of two machines are to be set Via a dialog object.

Create the following Frame:

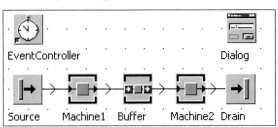

S. Bangsow: Manufacturing Simulation with Plant Simulation, Simtalk, pp. 253 – 272, 2010.
© Springer Berlin Heidelberg 2010

The dialog is to control the following settings:

Object	Attribute
Machine1	*ProcTime failed*
Machine2	*ProcTime failed*
Buffer	*Capacity*

Initially set the processing time of the machines to 1 minute. Insert a dialog object into the frame.

10.2.1 The Dialog Object

Double clicking or selecting Open on the context menu opens the dialog object.

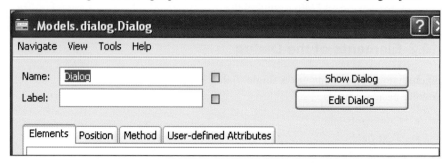

Clicking the button **SHOW DIALOG** allows you to view the dialog. Clicking **EDIT DIALOG** opens a dialog editor. You can arrange the individual elements of the dialog on the tab **ELEMENTS**. You can determine the position of the dialog box when it is called on the tab **POSITION**.

10.2.2 Insert Elements

Static text boxes explain the dialog. Use static text boxes as labels of the input text boxes and for general operating instructions. Select the item you want to insert on the context menu on the tab Elements.

Example: To enter the processing time of Machine1, you need a static text box (as identifier) and an edit text box (to enter text).

1. Click the right mouse button on the tab Elements, and select New Static Text Box.

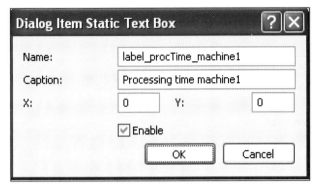

Name: *This name is the address of the text box.*

Caption: *This text is displayed on the dialog box.*

X/Y: *X and Y are the positions of the element in the dialog (column, row). The positions start at X = 0, Y = 0 (top left). You can set this quite easily afterwards by clicking the button Edit Dialog. Then, drag the fields to the correct position.*

2. Click again with the right mouse button on the tab Elements. Select New Edit Text box on the context menu: Enter the following data:

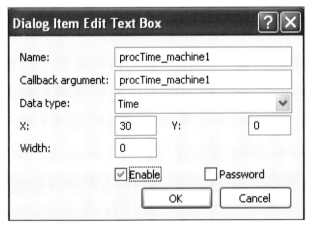

Callback argument: Using the callback argument allows you to access the field.

Data type: The setting in the Data type field restricts the possibilities of user input.

Enable: If you clear the checkmark, the item is disabled and no user input is accepted.

Password: Entries are masked.

You can double-click each element in the tab elements to open and edit it.

Click the button Show Dialog:

10.2.3 Callback Function

When you click OK, Cancel, or Apply, the dialog calls a method and passes a value (the callback argument), which makes it possible to recognize which button the user has clicked. The method is defined in the tab Method (default: self.callback). You can open the method by pressing F2.

There are predefined three arguments:

- Open
- Close
- Apply

OK calls the callback function twice: Apply and Close call it once each.

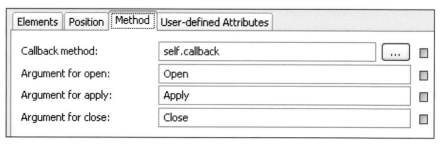

Within the callback function you need to program what is to happen when the user clicks on the respective buttons.

Example: When the user clicks the buttons in the dialog, the console should display a relevant message (to demonstrate). The callback function should have the following form:

```
(action : string)
is
do
    inspect action
    when "Open" then
        print  "Dialog open";
```

```
when "Apply" then
        print "Apply clicked";
when "Close" then
        print "OK or Cancel clicked";
end;
end;
```

10.2.4 The Static Text Box

Static text boxes are needed to display text in the dialog. You can modify the contents of the static text box at runtime. Use the method <path>.setCaption (<string1>, <string2>). The method setCaption sets the caption of the dialog element <string1> to the text <string2>.

Example (Method callback):

When you click Apply, the text of the static text box is to change.

```
(action : string)
is
do
    inspect action
    when "Open" then
    when "Apply" then
      dialog.setCaption("label_procTime_machine1",
      "Hello !");
    when "Close" then
      dialog.setCaption("label_procTime_machine1",
      "Processing time machine1");
    end;
end;
```

When you click Apply, the caption of the static text box is to change:

10.2.5 The Edit Text Box

The user can enter text into edit text boxes.

Important methods are

Method	Description
`<path>.setCaption(<string1>, <string2>)`	Set the contents of the text box `<string1>` to the new contents `<string2>`
`<path>.getValue(<string1>)`	Returns the contents of the text box `<string1>` (data type text).
`<path>.setSensitive(<string1>,<Boolean>)`	Activates /deactivates the element `<string1>`

Example: When opening the dialog, the edit text box should show the current processing time of the Machine1. Clicking Apply should set the processing time of Machine1 anew. For the next step, you need conversion functions: str_to_time (<string>) to convert text to the data type time, and to_str(<any>) to output any data type as text.

Callback function:

```
(action : string)
is
do
  inspect action
  when "Open" then
    --enter the processing time of Machine1
    --into the text box
    dialog.setCaption("procTime_machine1",
    to_str(machine1.procTime));
  when "Apply" then --new procTime machine1
    machine1.procTime:=
str_to_time(dialog.getValue("procTime_machine1"));
  when "Close" then
  end;
end;
```

10.2.6 Images in Dialogs

You can display images in the dialog, which you have previously defined in the icon editor as an icon of the dialog (Context menu – **EDIT ICONS** …).

Example: Insert a new icon in the report in the icon editor. Use an icon from the icon library (TOOLS – LOAD ICON).

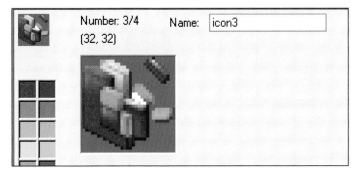

Select New Image from the context menu on the tab Elements. Enter the icon with the number from the icon editor (3 in our example).

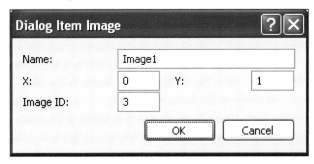

The image is displayed in the second row (Y=1) in the first column (X=0).

Methods:

Method	Description
<path>.setIcon(<string1>, <string2>)	Sets the image id of the image <string1> to <string2>.
<path>.getIcon(<string1>)	Returns the image id of the image <string1> as string.

10.2.7 Buttons

If the option SHOW DEFAULT BUTTONS on the tab ELEMENTS is selected, the dialog is displayed with three standard buttons.

You can also create a dialog with your own buttons.

Example 104: Error Dialog

Suppose you want to design your own message window. It will contain an error message and a symbol, furthermore an OK button which closes the window. Create a dialog (Error_Message), and insert an image (error_image) and a static text box (message).

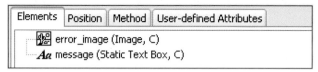

Clear the checkbox SHOW DEFAULT BUTTONS. Then add a button on the tab Elements. Enter the following settings:

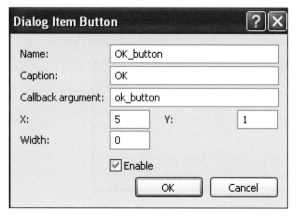

You have to program in the callback function the closing of the window. It could look as follows:

```
(action : string)
is
do
    inspect action
    when "ok_button" then
      error_Message.close(false);
    end;
end;
```

Some methods of the button:

Method	Description
`<path>.setCaption(<string1>, <string2>)`	Sets the caption of the button with the name `<string1>` on `<string2>`.
`<path>.setSensitive( <string1>,<Boolean>)`	Activates /deactivates the button `<string1>`

10.2.8 Radio Buttons

Radio buttons can only represent two values (true and false). Multiple option fields allow selection from a given set of options. A new selection deselects the previous selection. The assignment of the radio buttons to each other is set with a GroupID (integer). All fields with the same GroupID belong together.

Example: For Machine1, different states should be offered: operational, paused, and failed. The radio buttons have their own label (caption). Expand the dialog of the example of three radio buttons:

- ⦿ machine1_operational (Radio Button, C)
- ⦿ machine1_failed (Radio Button, C)
- ⦿ machine1_paused (Radio Button, C)

Settings of the radio buttons are: GroupID=0, callback arguments: machine1_operational, machine1_failed, Machine1_paused.

Labels:

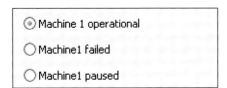

You must set up an area in the branch. Apply the callback function for each radio button in which you specify what is to happen if the option is selected. If you set the callback arguments up directly as a query (e.g., when "machine1_failed" then ...) in the callback function, the change will take place without the user having clicked Apply or OK (clicking the radio buttons call the callback function). The method *<path>.getCheckBox(<string>)* returns which radio button was selected:*

```
<path>.getCheckBox(<string>)
```

The return value has the data type boolean.

Method self.callback:

```
(action : string)
is
do
   inspect action
   when "Open" then
   when "Apply" then --new procTime machine1
      --get state of radio buttons and set state
      --of machine1
      machine1.failed:=
      dialog.getCheckBox("machine1_failed");
      machine1.pause:=
      dialog.getCheckBox("machine1_paused");
   when "Close" then
   end;
end;
```

When opening it, the dialog is to display the state of machine1. Set the value of the radio box with the method

```
<path>.setCheckBox(<string>,<boolean>);
```

Example: In the callback function above, you have to expand the branch for Open as follows:

```
---
when "Open" then
    dialog.setCheckBox("machine1_failed",machine1.failed);
    dialog.setCheckBox("machine1_paused",
                          machine1.pause);
    if not machine1.failed and not
       machine1.pause then
         dialog.setCheckBox("machine1_operational",
         true);
    end;
when "Apply" then
----
```

The radio button provides the following methods:

Method	Caption
<path>.setCaption(<string1>, <string2>)	Sets the caption of the radio button with the name <string1> on <string2>.
<path>.setSensitive(<string1>,<Boolean>)	Activates/deactivates the radio button <string1>

`<path>.setCheckBox(<` `string>, <boolean>);`	`Sets the status <boolean> of element` `<string>.`
`<path>.getCheckBox(<` `string>)`	`Returns the status of the element` `<string>`

10.2.9 Checkbox

The Checkbox can be either selected or cleared. The Checkbox represents two states (e.g., paused, not paused). The Checkbox provides the same methods as the radio button.

10.2.10 Drop-Down List Box and List Box

If you want to provide a great number of choices, then radio buttons require too much room in the dialog. For such cases, you can use (there are always some items visible in a list) drop-down list boxes (also list boxes, but there is only one entry visible; only for a selection, the list will be expanded).

Example 105: Statistics Dialog

A dialog is to display statistical data of different objects. Add a dialog to the Frame and name it statistics_dialog. Add the following elements to the dialog.

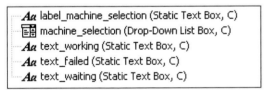

Captions:

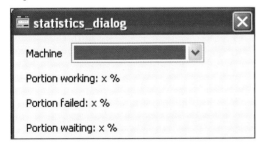

You can easily enter the values of the drop-down list into a table (click the button ITEMS in the properties dialog):

You can identify the selected entry with the method `<path>.getValue` `(<string>)` (pass the name of the drop-down list box). To access the corresponding objects (e.g., machine), you have to convert the string into an object (`str_to_obj(<string>)`).

Additional methods of the drop-down list box are

Method	Description
`<dialog>.setIndex` `(<string1>, <integer>)`	Selects the entry with the index `<integer>` in the dialog element with the name `<string1>`.
`<dialog>.getIndex(` `<string>)`	Determines the index of the selected list entry

An evaluation (e.g., after clicking the Apply button, or selecting another entry in the drop-down list) might look as follows:

```
(action : string)
is
  mach:object;
do
  inspect action
  when "Open" then
  when "Apply","maschine" then
    --get entry and read statistical data
    mach:=str_to_obj(statistics_dialog.getValue(
    "machine_selection"));
    --set captions of static text boxes
    statistics_dialog.setCaption("text_working",
    "Portion working:"+
    to_str(round(mach.statWorkingPortion*100,2))+
              " %");
    statistics_dialog.setCaption("text_failed",
    "Portion failed: " +
    to_str(round(mach.statFailPortion*100))+ " %");
    statistics_dialog.setCaption("text_waiting",
    "Portion waiting: " +
    to_str(round(mach.statWaitingPortion*100))+
              " %");
  when "Close" then
  end;
end;
```

10.2.11 List View

The list view displays the contents of a table in a dialog. The user can select a row from the table (the ListView returns the number of the selected row).

Example 106: Dialog Product Mix with Listview

A company manufactures three parts in a product mix. You are to simulate various mixes. You are to test a selection of mixes (a certain amount of part1, part2, and part3 each).

The proportion of part1, part2, and part3 is 1:3:5. First, create three entities (part1, part2, part3) in the class library. Create a table "mix" in the Frame with the following content:

	string 0	integer 1	integer 2	integer 3
string		part1	part2	part3
1	mix1	1	3	5
2	mix2	5	15	25
3	mix3	10	30	50
4	mix4	20	60	100
5	mix5	100	300	500

Dialog (mix_choice): Insert a static text box (as title) and a list view (mixtable).

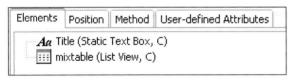

Setting ListView:

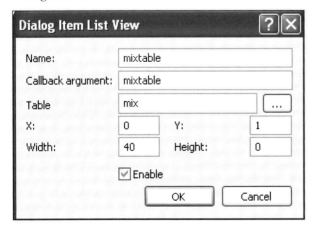

Dialog:

Methods of the ListView

Method	Description
`<path>.setTable(<string>, <object>)`	Sets the table of the element `<string>` to `<object>`
`<path>.setSensitive( <string1>,<Boolean>)`	Activates/deactivates the radio button `<string1>`
`<path>.setTableRow(<string>, <integer>)`	Sets the selected row in the ListView
`<path>.getTable(<string>)`	Returns the table name of the element `<string>`
`<path>.getTableRow(<string>)`	Returns the index of selected row (starting from 1)

Example: You need a production table (production) for the source (MU-selection: sequence cyclical).

	object 1	integer 2	string 3
string	MU	Number	Name
1	.MUs.part1	1	
2	.MUs.part2	3	
3	.MUs.part3	5	

When the user clicks Apply or OK, the number of parts from the selected mix in the rows of the table production will be transferred. After that, the source produces the new mix.

Callback function:

```
(action : string)
is
    row:integer;
do
```

```
    inspect action
    when "Open" then
    when "Apply" then
        -- TODO: add code for the "Apply" action here
        -- read index of selected row
        row:=mix_choice.getTableRow("mixtable");
        -- write selected product mix
        production[2,1]:=mix[1,row];
        production[2,2]:=mix[2,row];
        production[2,3]:=mix[3,row];
    when "Close" then
    end;
end;
```

10.2.12 Tab Control

If you have to arrange a number of elements on your dialog, presenting them on several tabs is helpful. All elements, which belong to a topic, are grouped on one tab. You first have to create a tab control and then add tabs to this element. New items are always created on the tab on which they are to be shown (context menu of the tab). The dialog structure is shown as a tree.

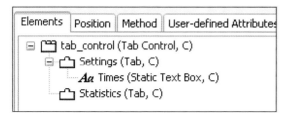

10.2.13 Group Box

The group box graphically groups elements. Create the elements within the group box using the context menu of the group box. The group box has a separate address (elements in the group start again at field position x = 0, y = 0). Elements of the group are displayed in the tree below the group box.

10.2.14 Menu and Menu Item

Menu items work just like buttons. When the user clicks a menu item, Plant Simulation passes the callback argument to the callback function. Within it you can evaluate the callback argument and initiate the necessary action.

Example 107: Dialog Menu

You are to control the simulation with a dialog. It is to be a menu in the dialog with the menu items: Start, Stop, Restart (stop, reset, init, start). Name the call-back arguments same as the menu items. First, insert the menu Simulation. Then create the menu items using the context menu of the menu.

Dialog:

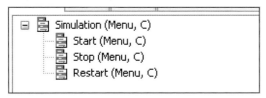

Callback method:

```
(action : string)
is
do
    inspect action
    when "Start" then
        eventController.start;
    when "Stop" then
        eventController.stop;
    when "Restart" then
        eventController.stop;
        eventController.reset;
        eventController.init;
        eventController.start;
    end;
end;
```

Methods of the menu:

Method	Description
`<path>.setCaption(<string1>, <string2>)`	Sets the caption of the menu/menu item with the name `<string1>` on `<string2>`.
`<path>.setSensitive (<string1>,<Boolean>)`	Activates/deactivates the menu/menu item `<string1>`

10.3 Accessing Dialogs

Important methods to access the dialogs are

Method	Description
`<path>.open`	`Shows the dialog`
`<path>.openDialog`	`Opens the dialog of the dialog`
`<path>.close`	`Hides the dialog`
`<path>.closeDialog`	`Closes the dialog window`

When you open the Frame, the dialog is to be shown automatically. The method must first show the Frame and then the dialog window.

```
is
do
   .models.frame.openDialog;
   dialog.open;
end;
```

The method will be allocated in the Frame with: TOOLS > SELECT CONTROLS …
Here you can define different actions and how they are triggered. The method is to be executed once the Frame is opened (Open).

10.4 Protection of Methods and Objects

You can quite easily hide the source code of methods from prying eyes. SELECT TOOLS > ENCRYPT in the method editor.

You will be prompted to enter a password. From then on you must enter this password before being able to edit the source code. To do so, select Tools > Decrypt. You can use this function, even for custom objects (e.g., Frames), whose content should not be seen and modified by everyone. It is easy to assign a method to the action Open the Frame, which opens the Frame dialog, e.g., only after entering a password (essentially you should encrypt the method in such a case).

Example 108: Protection of Frames

A Frame (frame1) is to be opened only after entering the password "Password". A method in the Frame controls the query of the password and opening of the dialog of the Frame.

Create a Frame protectFrame and insert a method (TOOLS > SELECT CONTROLS > OPEN).

```
is
   input:string;
   pass:string;
```

```
do
  pass:="xyz";
  input:=prompt("Password");
  if input = pass then
    .models.protectedFrame.openDialog;
  end;
end;
```

10.5 Validation User Input

If you use dialogs for user input it is important to validate the permissibility of the user input for the respective purpose (to avoid a runtime error). The validation has to take place prior to an assignment of the value to a property of an object (data type!).

10.5.1 Type Validation and Plausibility Check

You should first use all possibilities of the Edit Text box in the dialog (definition of the required data type). Plant Simulation then prevents the input of illegal characters. One way of validation is the following: Str_to_num returns the entered number; if text is entered, the return value is 0. You can use the return value 0 for a query.

Example 109: Type Validation

The user has to enter the size of a buffer into your dialog. If the user enters 0 or text, an error message will be displayed. The section of the call back method (Apply) might look as follows:

```
if str_to_num(dialog.getValue("capacity_buffer"))=0
then
    messagebox("Insert only values greater then 0!",
    1, 11);
else
    buffer.capacity:=str_to_num
    (dialog.getValue("capacity_buffer"));
end;
```

10.5.2 Message Box

You can display a message box using:

```
Messagebox("Message", integer buttons, integer i-
con);
```

Use the following values for buttons in Microsoft$^{©}$ Windows$^{©}$.

Value	Description
1	OK
3	OK, Cancel
10	Repeat, Cancel
48	Yes, No
50	Yes, No, Cancel

For the icon, you can pass the following values.

Value	Description	Icon
0	no icon	
1	Error	
2	question mark	
3	exclamation mark	
4	information	

If the user clicks one of the buttons, the method messagebox returns one of the following values:

- OK → 1
- Cancel→ 2
- Yes → 16
- No → 32

Note:

You can insert carriage returns and tabs into the text using the function Chr(ascii code). Enter Chr(9) for tabs and Chr(13) for carriage returns.
Sample: "text1"+Chr(13)+"text2"

10.6 HTML-Help

Plant Simulation provides a function to open a browser window and show a file in it. Syntax:

```
OpenHTMLWindow(<string location>, <string title>,
<integer x-Pos>, <integer y-Pos>, <integer width>,
<integer height>);
```

Example:

```
is
do
    OpenHTMLWindow("file:///c:/help.html", "Help",
    200,200,300,200);
end;
```

This opens a browser window (e.g., Internet Explorer). The HTML file will be loaded in this window.

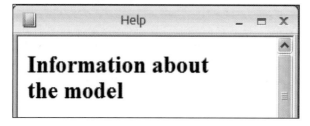

The HTML window can be closed with the command:

```
CloseHTMLWindow(string title);
```

Example:

```
is
do
  closeHTMLWindow("Help");
end;
```

11 Data Exchange

11.1 DDE with Plant Simulation

Dynamic data exchange (DDE) allows accessing another program. A Windows program that makes DDE functionality available connects to the system under a server name and provides various topics. You can establish a connection to this program by providing the specified server name and a valid theme. You may use the established channel for certain transactions related to data (Items) that are supported by the server. All DDE transactions are realized by SimTalk calls in Plant Simulation. You can call any method in Plant Simulation with DDE, including the methods, which you have defined.

Important: All participating programs must be started!

11.1.1 Read Plant Simulation Data in Microsoft Excel

Example 110: Data Exchange DDE Excel

The following examples are programmed in VBA, in principle this works with Excel, Project, Word, and Access. Insert a variable (name: variable, data type: integer, value: 247985) into a Frame (.Models.Frame). This value is to be read in Excel. Start Excel. Open the Visual Basic Editor. Insert a new module (Insert – Module) and a new procedure (sampleDDE) in the module:

```
Public Sub sampleDDE()
Dim channel As Long
Dim value As Variant
'establish channel
channel = DDEInitiate("eM-Plant", "Data")
read the value
value = DDERequest(channel, _
".Models.Frame.Variable")
'write the value into the table
Tabelle1.Range("A2").value = value(1)
'close channel
DDETerminate (channel)
End Sub
```

To work with DDE, you need the following commands:

S. Bangsow: Manufacturing Simulation with Plant Simulation, Simtalk, pp. 273–287, 2010.
© Springer Berlin Heidelberg 2010

VBA/ Simtalk -Method	*Description*
DDEInitiate(<string1>, <string2>)	Opens a channel. You have to pass the name of the application (<string1>, (in Plant Simulation 9 this still is eM-Plant) and a topic (<string2>). Plant Simulation supports the following topics: System (You can request information about Plant Simulation, you can call SimTalk methods) Data (Values of global variables you can read and write) Info (Version of Plant Simulation, name of the current frame, states of the frame you can read)
DDERequest(<integer>, <string>)	Requests information from an application. You need to pass the channel number (<integer>) and an item (<string>). The passed element (item) will be queried. (Note: It must be addressed absolutely.)
DDETerminate(<integer>)	Closes the channel (<integer>).
DDEPoke(<integer>, <string1>, <variant>)	Writes data (variant>) into the connected application (channel <integer>, element <string>).
DDEExecute(<integer>, <string>)	Sends a command (<string>) to the connected application (<integer> channel.

11.1.2 Excel Data Import in Plant Simulation

Plant Simulation can also read data from DDE-enabled programs. Regarding Excel, a few requirements must be met:

- The Excel file must be opened.
- You have to specify the worksheet in Excel when opening the connection.
- All data will be transferred as a string.

Example 111: Data Exchange, Importing a Working Plan from Excel

You are to import a work plan from Excel into Plant Simulation. The work plan should have, for example, the following form (insert a table „work plan" into the Frame in Plant Simulation).

	string 1	string 2	string 3	string 4
1	Machine	Processing time	Successor	
2	M1	10	M2	
3	M2	20	M4	

Then, create a table in Excel, and enter the following data:

	A	B	C
1	Machine	Processing time	Successor
2	M1	10	M2
3	M2	20	M4
4	M4	30	M6
5	M6	10	Drain

You are to transfer the data from Excel to Plant Simulation. The following example loads all data in the Excel-Table2 into a Plant Simulation table (working_plan).

Method: load_workingplan

```
is
   value, adress:string;
   channel, row, column:integer;
   colNext, rowNext:boolean;
do
   --establish connection
   channel:=DDEConnect("Excel","Table2");
   --column by column, until no more values available
   row:=1;
   column:=1;
   rowNext:=true;
   while(rowNext) loop
   --start from address 1,1 column by column
   column:=1;
   colNext:=true;
   --no value, stop both loops
   adress:="R"+to_str(row)+"C"+to_str(column);
     value:=ddeRequest(channel,adress);
     --returns additional a line break
     if value = chr(13)+chr(10) then
        rowNext:=false;
        colNext:=false;
     else
        while (colNext) loop
          adress:="R"+to_str(row)+"C"+to_str(column);
```

```
        value:=ddeRequest(channel,adress);
        if value = chr(13)+chr(10) then
          colNext:=false;
        else
          working_plan[column,row]:=
          omit(value,strlen(value)-1,2);
          column:=column+1;
        end;
      end;
    row:=row+1;
    end;
  end;
  ddedisconnect(channel);
end;
```

Note:
Check to see what Excel returns, if you read the value of an empty cell. In our example, Excel 2007 returns a carriage return (ASCII 10 and 13) at the end of each value. You have to remove this carriage return before you write the value into the Plant Simulation table. Within the method load_workingPlan we used the method `omit (<string>, <integer>, <integer>)`.

11.1.3 Plant Simulation Remote Control

You can call Plant Simulation commands with the function DDEExecute (channel, command) in Excel.

Example 112: DDE Remote Control

Suppose you want to control the Frame .Models.Frame from an Excel file. The simulation is to start and stop with buttons in Excel. Add two buttons (ActivX controls) to an Excel spreadsheet.

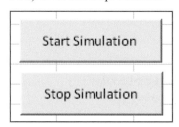

Double-clicking opens a module sheet (VBA). Event handling for starting the simulation might look as follows.

```
Private Sub CommandButton1_Click()
Dim channel As Long
```

```
'Establish connection
channel = DDEInitiate("eM-Plant", "System")
'Execute command
DDEExecute                            channel,
".models.frame.eventController.start"
'close connection
DDETerminate (kanal) 'kanal wird geschlossen
End Sub
```

The buttons are activated after terminating design mode. Click the Development

tools toolbar in Excel on the icon ![icon] .

11.1.4 DDE Hotlinks

A simple way of exchanging data is using links. The values of the links are automatically updated. Within DDE these links are called hotlinks. To make it work, you must first enable the corresponding global variable for the hotlink.

Example 113: DDE Hotlinks

Insert a global variable named variable into a Frame. The variable is to indicate the number of completed parts. You need a method (e.g., entrance control drain). The method might look as follows:

```
is
do
    variable:=drain.statNumIn;
end;
```

You have to enable the variable for DDE Hotlinks. Select the option Support DDE Hotlinks in the dialog of the variable on the tab Communication.

Value	Display	Statistics	Communication	Com
☑ Support DDE hotlinks			⊟	

In Excel, you can now set a hotlink to the variable in a table cell:

='eM-Plant'|Data!'.models.frame.variable'				
E	F	G	H	
	output	2345		

Once the value of the variable changes, this will update the value of the cell automatically.

11.2 The File Interface

Using the file interface, you can access text files to read, delete, and overwrite their contents.

Example 114: File Interface

For the example you need an object of type Comment, one file interface, and two Method objects.

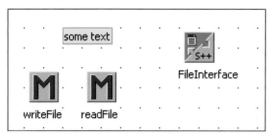

Type some text into the comment of the Comment object. This text is to be written into a file. In the Comment object disable the option:

Create a file fileinterface.txt in the same folder as the Plant Simulation file. You first have to set up the file interface. Open the object FileInterface, and select or enter the following settings:

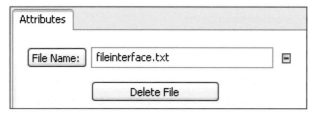

Method writeFile:

```
is
do
    fileInterface.open;
    fileInterface.writeln(comment.cont);
    fileInterface.close;
end;
```

The content of the Comment object will be written to the file.

The contents of the file are now to be displayed in the console.

Method: readFile

```
is
   i:integer;
do
   --writes content of the textfile in the console
   fileInterface.open;
   i:=1;
   fileInterface.gotoline(i);
   while not fileInterface.eof loop
     print fileInterface.readln;
     i:=i+1;
     fileInterface.gotoline(i);
   end;
   fileInterface.close;
end;
```

Attributes and methods of the FileInterface:

Method/Attribute	Description
<path>.fileName	Sets/writes the filename of the FileInterface.
<path>.remove	Deletes the specified file.
<path>.goToLine(<integer)	Sets the pointer of the file on the specified line. In this way, you can write and read line by line.
<path>.eof	Returns true, if the end of file is reached.
<path>.writeLn (<string>)	Overwrites the line at the current line position with the given content.
<path>.readLn	Returns the content of the actual line as string.

11.3 The ODBC Interface

The ODBC interface allows to access ODBC data sources. You can, for example, use the ODBC interface for reading data from a database, for modifying data in databases, and for entering data from Plant Simulation into a database. You must first add the object ODBC to the Class Library (for this you need the Plant Simulation interface package). From the Plant Simulation menu bar select: FILE – MANAGE CLASS LIBRARY

Look for the ODBC interface in the branch INFORMATIONFLOW:

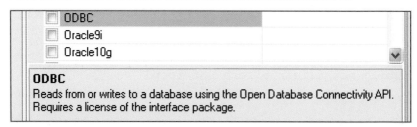

Select ODBC and confirm your selection with OK.

11.3.1 Setup an ODBC Data Source

Before you can use the ODBC Interface, you must create a database and set the database up as an ODBC data source.

Example 115: ODBC

You are to simulate two machines and one buffer. Create the following Frame:

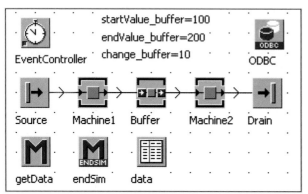

The Source produces one part (blocking) every minute. You are to save the settings and the results of the simulation into a database. Insert a database into Access and save the database under productiondatabase (.mdb). The database will initially contain two tables:

Table Elements:

Field name	Data type	Commit
ID	Auto value	Primary key
Name	Text	Maximum 50 characters
Processing_time	Number	Long integer (seconds)
Availability	Number	Double
MTTR	Number	Long integer (seconds)
Capacity	Number	Long integer (default 1)

Table simulation_runs:

Throughput, average exit interval (cycle time), and output as a function of buffer size are to be recorded.

Field name	Data type	Commit
ID	Auto value	Primary key
Buffer_capacity	Number	Long integer
Throughput	Text	
Exit_interval	Text	
Output	Number	Long Integer

Type the following data in the table elements:

ID	Name	Processing_time	Availability	MTTR	Capacity
1	Machine1	50	75	18000	1
2	Buffer	0	100	0	100
3	Machine2	45	50	36000	1

You have to set up the database as ODBC data source. Under MS Windows XP, you select Start – Control Panel – Administrative Tools – Data sources (ODBC). Click the tab System DSN, and then click Add. Select the ACCESS- database driver and click Finish. In the next window, type in a name for the Data source (production_database) and select the database. The database is accessed on the data source name (DSN).

11.3.2 Read Data from a Database

Open the ODBC object. Enter the name of the ODBC data source into the field database.

```
.Models.odbc_interface.ODBC

Name:       ODBC                         □

Label:                                   □


Database:   production_database
```

Confirm your changes. No other settings are required in this window.

Example: You are to read all data from the database table elements into the table data.

Method getData:

```
is
   sql:string;
do
   --login
   ODBC.login("production_database","","");
   --form sql-command
   sql:="SELECT * FROM Elements";
   -- send sql command,
   --Plant Simulation writes the result
   --into the table data
   ODBC.sql(data,sql);
   -- logout
   ODBC.logout;
end;
```

Execute the method. Plant Simulation saves the query results in the table data. You can read the data from there.

	integer 1	string 2	integer 3	real 4	integer 5	integer 6
string	id	Name	Processing_time	Availability	MTTR	Capacity
1	1	Machine1	50	75.000000	18000	1
2	2	Buffer	0	100.000000	0	100
3	3	Machine2	45	50.000000	36000	1
4						

The setting of the data using the table values may now look like this:

Method getData:

```
is
  sql:string;
do
  --login
  ODBC.login("production_database","","");
  --form sql-command
  sql:="SELECT * FROM Elements";
  -- send sql command, Plant Simulation writes
  -- the result
  --into the table data
  ODBC.sql(data,sql);
  -- logout
  ODBC.logout;
  -- set values
  for i:=1 to data.Ydim loop
    element:=str_to_obj(data[2,i]);
    element.procTime:=data[3,i];
    element.failures.failure.availability:=
    data[4,i];
    element.failures.failure.mttr:=data[5,i];
    element.capacity:=data[6,i];
  next;
end;
```

Note:
Before executing the method, you must insert a failure to all objects (without any settings). The default name of the failure is "failure". Then the method works without error messages.

11.3.3 Write Data in a Database

Writing data in a database is analogous to reading data. You will establish a connection to the database, send an SQL statement, and close the connection after completing working with the database. For entering new records, you use the SQL command INSERT; for modifying existing records UPDATE.

Example 116: ODBC – Write Simulation Results into a Database

Suppose you want to automatically execute a series of simulation runs. The aim is to increase the size of the buffer by the value startvalue_buffer up to a value end-value_buffer to a size change_buffer. Each variant will be simulated (setting in the EventController). After completing a run (endSim), the values are to be entered into a new row in the table simulation_runs.

Insert the global variables into the Frame, and enter the following initial values:

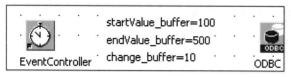

startValue_buffer=100

endValue_buffer=500

change_buffer=10

EventController

ODBC

For entering the data into the table, you will use the SQL command INSERT. The syntax of this command is:

```
INSERT INTO table (field1, field2, field3 …) VALUES
(value1, value2, value3 …).
```

Strings must be specified in the SQL statement in single quotation marks.

The Method init might look as follows:

```
is
   sql:string;
do
   --daten auslesen
   sql:="INSERT into simulation_runs "+
       "(Buffer_capacity,"+
       "Throughput,Exit_interval,Output) Values ("+
       to_str(buffer.capacity)+","+
       "'"+to_str(drain.statavgLifeSpan)+"',"+
       "'"+to_str(drain.statavgExitInterval)+"',"+
       to_str(drain.statnumOut)+")";
   ODBC.login("production_database","","");
   ODBC.sql(sql);
   ODBC.logout;
   --if buffer capacity not endvalue_buffer increase
   -- and restart the simulation
   if buffer.capacity < endvalue_buffer then
      buffer.capacity:=buffer.capacity+change_buffer;
      eventController.reset;
      eventController.start;
   end;
end;
```

11.3.4 Delete Data in a Database Table

You can delete values from database tables with the SQL command DELETE. If you do not restrict deletion by a WHERE clause, all data in the table will be deleted.

Example 117: ODBC – Delete Data

You want to delete all data in the table simulation_runs with a SimTalk method. The SQL command is

```
DELETE FROM simulation_runs.
```

The method deleteAllRuns might look as follows:

```
is
  sql:string;
do
  sql:="DELETE FROM simulation_runs";
  ODBC.login("production_database","","");
  ODBC.sql(sql);
  ODBC.logout;
  --buffer capacity reset
  buffer.capacity:=startvalue_buffer;
end;
```

11.3.5 SQL Commands

The database interprets all text information except the commands in SQL, which you define, as a table or column name. You must set all text values within a SQL statement within single quotation marks. The databases support different ranges of SQL statements; the selection below is the lowest common denominator. A good documentation about the entire range of SQL in its current version can be found on this Web site:

http://dev.mysql.com/doc/refman/6.0/en/sql-syntax.html

Access only supports a small fraction of the currently valid SQL syntax.

11.3.5.1 SELECT

```
SELECT  [DISTINCT | ALL]
Select_columns,...
  [FROM table
  [WHERE where_definition]
  [ORDER BY { column number | col. name | formula }
  [ASC / DESC]
```

A general query has the following form:

```
SELECT column FROM tableName WHERE column="value"
```

To show all columns:

```
SELECT * FROM tableName WHERE column="value"
```

To show all columns and all records:

```
SELECT * FROM tableName
```

If you want to sort the results of a query, you can insert an ORDER BY clause. You can specify the column position (starting with 1), column names, or an arithmetic expression for calculating the column position. The default sort order is ascending (ASC), if you want to sort in descending order, you must use DESC.

Example: You want display all records from the table elements order by name. The SQL command must look as follows:

```
SELECT * FROM Elements ORDER BY Name
```

Often it is expected that a word will be found, even if it is contained within another word or string. Sample: If you search for "Machine", both machines should be found. In SQL, the LIKE expression is used for this purpose. LIKE compares two strings. If they match, the record is included in the search results. Instead of using complete strings to compare, you can also use the so-called wild cards:

- _ one character
- % for an indefinite number of characters

Example:

```
SELECT * FROM Elements WHERE Name LIKE 'Machine%'
ORDER BY Name
```

11.3.5.2 INSERT (Insert New Records)

```
INSERT    [INTO] table [(column names,…)]
          VALUES (value1,…)
```

The simplest form is

```
INSERT INTO tableName VALUES
(value1,'value2',value_n)
```

The values in parentheses must form a complete data set (number and data types), otherwise, an error occurs!

11.3.5.3 UPDATE (Change Data)

```
UPDATE Table_name
       SET column_name1=value1, [column_name2=value2, …]
       [WHERE where_definition]
```

UPDATE updates columns in table rows with new values. The SET clause determines which columns are to be changed and which value they receive. The WHERE clause, if present, determines which rows will be changed. If it is missing, all rows will be changed.

Example: You want to change the processing time of Machine1 to 100 seconds.

SQL expression:

```
UPDATE Elements SET Processing_time=100 WHERE
Name='Machine1'
```

11.3.5.4 DELETE

```
DELETE FROM tbl_name
     [WHERE where_definition]
```

DELETE deletes rows from the table tbl_name, which meet the condition in the where_definition and returns the number of deleted rows. If you use the DELETE statement without a WHERE clause, all rows in the table will be deleted.

Example: You want to delete the Buffer in the table Elements:

```
DELETE FROM Elements WHERE Name='Buffer'
```

12 Plant Simulation 3D

12.1 Sample Project

Plant Simulation supports modeling and simulating models in virtual space. The 2D model is assigned a 3D model, which is controlled by the 2D model (corresponding models). All changes in 2D/3D model have an impact on the corresponding model in the other part of the program.

Example 118: Plant Simulation 3D

You are to simulate two machines which are connected with a conveyor belt. The conveyor belt is divided into three segments of 1 meter each. The conveyor speed is 0.5 m/min, the processing time of the machines is 1 minute, and the source generates one part every minute. Create the following Frame:

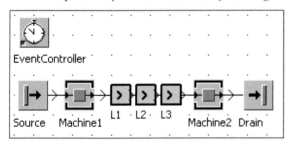

Now start Plant Simulation 3D: 3D – Start 3D viewer.

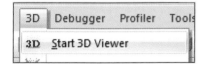

Now you have two corresponding models. The preliminary result is a 3D model with standard symbols.

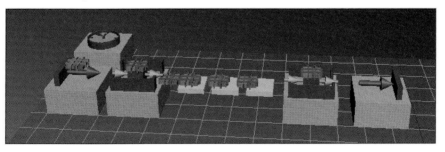

S. Bangsow: Manufacturing Simulation with Plant Simulation, Simtalk, pp. 289 – 291, 2010.
© Springer Berlin Heidelberg 2010

12.2 Views and Move in Plant Simulation 3D

Corresponding to the size of the corresponding 2D simulation, Plant Simulation generates a base plate in the 3D model. As a 2D reference level (Z = 0), the base plate facilitates the orientation and navigation within 3D space. The base plate has a grid on which you can align your objects. You can display or hide the grid button:

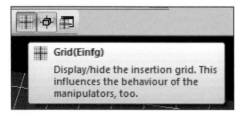

Click Modes – Motion for moving the actual scene. Plant Simulation offers the following ways to move the scene:

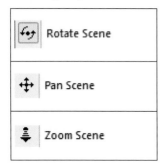

You can use the icons with the red dot to manipulate individual objects.

If you are lost, and your Frame is no longer shown on the screen, you can use the command View > View All.

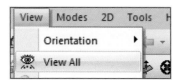

12.3 Control the Simulation in Plant Simulation 3D

You can start the EventController with . In addition, you can find buttons to reset, start, and stop the simulation on the toolbar.

Symbol	Function
▷	Starts and stops the Simulation
◄◄◄	Stop + Reset
▷▷	Starts without MU-animation
🖌	Deletes all MUs

Index

Printing and Binding: Stürtz GmbH, Würzburg